BIOCHEMISTRY RESEARCH TRENDS

BIOCHEMISTRY RESEARCHER BIOGRAPHICAL SKETCHES AND RESEARCH SUMMARIES

BIOCHEMISTRY RESEARCH TRENDS

Additional books in this series can be found on Nova's website under the Series tab.

Additional e-books in this series can be found on Nova's website under the e-books tab.

BIOCHEMISTRY RESEARCH TRENDS

BIOCHEMISTRY RESEARCHER BIOGRAPHICAL SKETCHES AND RESEARCH SUMMARIES

JAMES T. WALKER
AND
CATHERINE M. JONES
EDITORS

New York

For permission to use material from this book please contact us:
Telephone 631-231-7269; Fax 631-231-8175
Web Site: http://www.novapublishers.com

Library of Congress Cataloging-in-Publication Data

Biochemistry researcher biographical sketches and research summaries / [edited by] James T. Walker and Catherine M. Jones.
p. cm.
ISBN 978-1-62081-308-9 (hardcover)
1. Biochemistry--Research--Abstracts. 2. Biochemists--Directories. I. Walker, James T., 1962- II. Jones, Catherine M. (Catherine Marie), 1976-
QH345.B5275 2011
572.025--dc23
2012007382

Published by Nova Science Publishers, Inc. † New York

Contents

Preface

This new book compiles biographical sketches of top professionals in the field of biochemistry research, as well as research summaries from a number of different focuses in this important field.

Part 1 - Research Biographies

In: Biochemistry Researcher Biographical Sketches ... ISBN: 978-1-62081-308-9
Editors: James T. Walker and Catherine M. Jones

Chapter 1

Cristina Andres-Lacueva

Affiliation:

Nutrition and Food Science Department, XaRTA-INSA. Pharmacy Faculty, University of Barcelona, Spain.

Address:

Pharmacy Faculty, University of Barcelona. Av. Joan XXIII s/n, 08028 Barcelona, Spain.

Date of Birth:

15/01/1969

Education:

PhD in Pharmaceutical Sciences

Research and Professional Experience:

2009: Visiting Scientist in the University of Perurgia. Department of Clinical and Experimental Medicine, Institute of Gerontology and Geriatrics.

2006: Visiting Scientist in the INRA – Institute National de Recherche Agronomique, France (3 months).

2003 – Present: Regular scientific External of the European Commission in the Evaluation committee of the Sixth and Seventh Framework Program. Different calls: "Fork to Farm:

Food, Nutrition and Well being" of the theme "Food, Agriculture and Fisheries, and Biotechnology"; Research for SME (Small & Medium Enterprise) Associations.

2003: Scientific Expert in the PHARE-TWINNING (RO00/IB/AG-04): "Harmonanisation of Legislation Strengthening the Capacity to Manage the acquis on Wine", financed by European Union programme PHARE. Project carried out by the Romanian Ministry of Agriculture, Food and Forest and the Ministry of Agriculture, Cattle Rising and Rural Development of Regional Government of La Rioja, Spain.

2003 – Present: Regular Expert Reviewer of the National Association for the Evaluation in Spain (ANEP).

2002-2004: Professor of the Master and Posgrade Distance in Nutrition and Food Sciences. Institute Micromat. University of Barcelona.

2002-2003: Responsable Professor "Nutrition and Toxicology" from the Master in Chemistry and Food ingeniering. Instituto Químico de Sarriá. University Ramon Llull. 2002-2003: Visiting Professor of Introduction to the Nutrition. Escuela Universitaria de Ingeniería Técnica Del Máster INEA en calidad total en la Industria Agroalimentaria Agrícola INEA (Valladolid).

2002-2005: Scientist "Ramon y Cajal" from the Ministry of Science and Technology of Spain. University of Barcelona.

2000-2002: Postdoctoral Fellowship and Postdoctoral Associate. Neurosciences Laboratory. USDA – Human Nutrition Research Center on Aging at Tufts University. Boston, Massachusetts.

1999 -2002: Assistant Professor. University of Barcelona, Dept. Nutrition and Food Sciences.

05/97-09/97: Research Scientist. Istituto Agrario San Michele all/Adige. Trento, Italy. Chemistry of Food and Beverages Laboratory

Professional Appointments:

Associate Professor Qualified in Nutrition and Food Science Area (June, 2005)

Leader of the Metabolomic and Biomarkers Group on Nutritional Studies at the Department of Nutrition and Food Science, Pharmacy School, University of Barcelona.

Accreditation of Advanced Research by AQU- Catalonia (September, 2009)

Publications Last 3 Years:

1. Darias-Martín, J.J.; Andrés-Lacueva, C.; Díaz-Romero, C.; Lamuela-Raventós, R.M. Phenolic profile in varietal white wines made in the Canary Islands. *Eur Food Res Technol.* 2008;226(4):871-876.
2. Andrés-Lacueva, C.; Monagas, M.; Khan, N.; Izquierdo-Pulido, M.; Urpí-Sardà, M.; Permanyer, J.; Lamuela-Raventós, R.M. Flavanol and flavonol contents of cocoa powder products: influence of the manufacturing process. *J Agric Food Chem.* 2008;56(9):3111-3117.
3. Cherubini, A.; Ruggiero, C.; Morand, C.; Lattanzio, F.; Dell'aquila, G.; Zuliani, G.; Di Iorio, A.; Andrés-Lacueva, C. Dietary antioxidants as potential pharmacological agents for ischemic stroke. *Curr Med Chem.* 2008;15(12):1236-1248.

4. Zamora-Ros, R; Andrés-Lacueva, C.; Lamuela-Raventós, R.M.; Berenguer, T.; Jakszyn, P.; Martínez, C.; Sánchez, M.J.; Navarro, C.; Chirlaque, M.D.; Tormo, M.J.; Quirós, J.R.; Amiano, P.; Dorronsoro, M.; Larrañaga, N.; Barricarte, A.; Ardanaz, E.; González, C.A. Concentrations of resveratrol and derivatives in foods and estimation of dietary intake in a Spanish population: European Prospective Investigation into Cancer and Nutrition (EPIC)-Spain cohort. *Br J Nutr.* 2008; 100(1):188-196.
5. Mata-Bilbao, M.L.; Andrés-Lacueva, C.; Roura, E.; Jáuregui, O.; Escribano, E.; Torre, C.; Lamuela-Raventós, R.M. Absorption and pharmacokinetics of green tea catechins in beagles. *Br J Nutr.* 2008;100(3):496-502.
6. Roura, E.; Andrés-Lacueva, C.; Mata-Bilbao, M.L.; Estruch, R.; Izquierdo-Pulido, M.; Lamuela-Raventós, R.M. The effects of milk as a food matrix for polyphenols on the excretion profile of cocoa (-)-epicatechin metabolites in healthy human subjects. *Br J Nutr.* 2008;100(4):846-851.
7. Lamuela-Raventos, R.M.; Roura, E.; Andres-Lacueva, C.; Estruch, R. Bioavailability of flavonoids. *Cerevisia.* 2009;34(1):26-30.
8. Lamuela-Raventos, R.M.; Medina-Remón, A.; Andres-Lacueva, C.; Estruch, R. El vino sólo se disfruta con moderación. *FIVIN difusión.* 2009.
9. Zamora-Ros, R.; Lamuela-Raventós, R.M.; Estruch, R.; Andrés-Lacueva, C. Resveratrol, a new biomarker of moderate wine intake? *Br J Nutr.* 2009;101(1):148.
10. Abbatecola, A.M.; Cherubini, A.; Guralnik, J.M.; Andrés-Lacueva, C.; Ruggiero, C.; Maggio, M.; Bandinella, S.; Paolisso, G.; Ferrucci, L. Plasma polyunsaturated fatty acids and age-related physical performance decline. *Rejuv Res.* 2009; 12(1):25-32.
11. Medina-Remón, A.; Barrionuevo-González, A.; Zamora-Ros, R.; Andrés-Lacueva, C.; Estruch, R.; Martínez-González, M.A.; Diez-Espino, J.; Lamuela-Raventós, R. M. Rapid Folin-Ciocalteu method using microtiter 96-well plate cartridges for solid phase extraction to assess urinary total phenolic compounds, as a biomarker of total polyphenols intake. *Anal Chim Acta.* 2009; 634(1):54-60.
12. Hussein, L.; Medina, A.; Barrionuevo, A.; Lamuela-Raventós, R.M.; Andrés-Lacueva, C. Normal distribution of urinary polyphenol excretion among Egyptian males 7-14 years old and changes following nutritional intervention with tomato juice (Lycopersicon esculentum). *Int J Food Sci Technol.* 2009;60(4):302-311.
13. Zamora-Ros, R.; Urpí-Sardà, M.; Lamuela-Raventós, R. M.; Estruch, R.; Martínez-González, M.A.; Bulló, M.; Arós, F.; Cherubini, A.; Andrés-Lacueva, C. Resveratrol metabolites in urine as a biomarker of wine intake in free-living subjects: The PREDIMED Study. *Free Radic Biol Med.* 2009; 46(12):1562-1566.
14. Urpí-Sardà, M.; Monagas, M.; Khan, N.; Lamuela-Raventós, R.M.; Santos-Buelga, C.; Sacanella, E.; Castell, M.; Permanyer, J.; Andrés-Lacueva, C. Epicatechin, procyanidins, and phenolic microbial metabolites after cocoa intake in humans and rats. *Anal Bioanal Chem.* 2009; 394(6):1545-1556.
15. Monagas, M.; Khan, N.; Andrés-Lacueva, C.; Urpí-Sardà, M.; Vàzquez-Agell, M.; Lamuela-Raventós, R.M.; Estruch, R. Dihydroxylated phenolic acids derived from microbial metabolism reduce lipopolysaccharide-stimulated cytokine secretion by human peripheral blood mononuclear cells. *Br J Nutr.*2009;102(2):201-206.
16. Guillén, M.; Corella, D.; Portolés, O.; Estruch, R.; Andrés-Lacueva, C. Efectos del vino tingo sobre la salud: papel de los polifenoles (resveratrol) y etanol. *VEREMA difusión.* 2009.

17. Urpí-Sardà, M.; Monagas, M.; Khan, N.; Llorach, R.; Lamuela-Raventós, R.M.; Jauregui, O.; Estruch, R.; Izquierdo-Pulido, M.; Andrés-Lacueva, C. Targeted metabolic profiling of phenolics in urine and plasma after regular consumption of cocoa by liquid chromatography-tandem mass spectrometry. *J Chromatogr A*. 2009; 1216(43):7258-7267.
18. Llorach, R.; Urpí-Sardà, M.; Jauregui, O.; Monagas, M.; Andrés-Lacueva, C. An LC-MS-Based metabolomics approach for exploring urinary metabolome modifications after cocoa consumption. *J Proteome Res*. 2009;8(11):5060-5068.
19. Monagas, M.; Khan, N.; Andrés-Lacueva, C.; Casas, R.; Urpí-Sardà, M.; Llorach, R.; Lamuela-Raventós, R.M.; Estruch, R. Effect of cocoa powder on the modulation of inflammatory biomarkers in patients at high-risk of cardiovascular disease. *Am J Clin Nutr*. 2009;90(5):1144-1150.
20. Urpí-Sardà, M.; Garrido, I.; Monagas, M.; Gómez-Cordovés, C.; Medina-Remon, A.; Andrés-Lacueva, C.; Bartolomé, B. Profile of plasma and urine metabolites after the intake of almond [Prunus dulcis (Mill.) DA Webb] polyphenols in humans. *J Agric Food Chem*. 2009;57(21):10134-10142.
21. Llorach, R.; Andrés-Lacueva, C. Metabolomics of cocoa consumption. *Bioanalysis, News & Analysis*. 2009;1(9):1518.
22. Ruggiero, C.; Lattanzio, F.; Lauretani, F.; Gasperini, B.; Andrés-Lacueva, C.; Cherubini, A. Omega-3 polyunsaturated fatty acids and immune-mediated diseases: inflammatory bowel disease and rheumatoid arthritis. *Curr Pharm Design*. 2009;15(36):4135-4148.
23. Llorach-Asunción, R.; Jauregui, O.; Urpí-Sardà, M.; Andrés-Lacueva, C. Methodological aspects for metabolome visualization and characterization: a metabolomic evaluation of the 24 h evolution of human urine after cocoa powder consumption. *J Pharm Biomed Anal*. 2010;51(2):373-381.
24. Andrés-Lacueva, C.; Zamora-Ros, R. Wanted: specific nutritional biomarkers for food consumption for the study of its protective role in health. *Br J Nutr*. 2010;103(3):307-308.
25. Zamora-Ros, R.; Andrés-Lacueva, C.; Lamuela-Raventós, R. M.; Berenguer, T.; Jakszyn, P.; Barricarte, A.; Ardanaz, E.; Amiano, P.; Dorronsoro, M.; Larrañaga, N.; Martínez, C.; Sánchez, M. J.; Navarro, C.; Chirlaque, M. D.; Tormo, M. J.; Ramón Quirós, J.; González, C.A. Estimation of dietary sources and flavonoid intake in a Spanish adult population (EPIC-Spain). *J Am Diet Assoc*. 2010;110(3):390-398.
26. Llorach, R.; Urpí-Sardà, M.; Rotchés-Ribalta, M.; Rabassa, M.; Andrés-Lacueva, C. Resveratrol: from dietary intake to promising therapeutic molecule. *Agro Food Ind Hi-Tech*. 2010; 21(2):42-44.
27. Urpí-Sardà, M.; Llorach, R.; Khan, N.; Monagas, M.; Rotchés-Ribalta, M.; Lamuela-Raventós, R.M.; Estruch, R.; Tinahones, F.J.; Andrés-Lacueva, C. Effect of milk on the urinary excretion of microbial phenolic acids after cocoa powder consumption in humans. *J Agric Food Chem*. 2010;58(8):4706-4711.
28. Urpí-Sardà, M.; Ramiro-Puig, E.; Khan, N.; Ramos-Romero, S.; Llorach, R.; Castell, M.; Gonzalez-Manzano, S.; Santos-Buelga C., Andrés-Lacueva, C. Distribution of epicatechin metabolites in lymphoid tissues and testes of young rats with a cocoa-enriched diet. *Br J Nutr*. 2010;103(10):1393-1397.

29. Khan, N.; Monagas, M.; Llorach, R.; Urpí-Sardà, M.; Rabassa, M.; Estruch, R.; Andrés-Lacueva, C. Targeted and metabolomic study of biomarkers of cocoa powder consumption: effects on inflammatory biomarkers in patients at high risk of cardiovascular disease. *Agro Food Ind Hi-Tech*. 2010;21(3):54-57.
30. Ortuño, J.; Covas, M.I.; Farré, M.; Pujadas, M.; Fitó, M.; Khymenets, O.; Andrés-Lacueva, C.; Roset, P.; Joglar, J.; Lamuela-Raventós, R.M.; de la Torre, R. Matrix effects on the bioavailability of resveratrol in humans. *Food Chem*. 2010;120(4):1123-1130.
31. Monagas, M.; Urpi-Sarda, M.; Sanchez-Patan, F.; Llorach, R.; Garrido, I.; Gomez-Cordoves, C.; Andres-Lacueva, C.; Bartolome, B. Insights into the metabolism and microbial biotransformation of dietary flavan-3-ols and the bioactivity of their metabolites. *Food and Function*. 2010;1(3):233-253.
32. Bartolomé, B.; Monagas, M.; Garrido, I.; Gómez-Cordovés, C.; Martín-Álvarez, PJ.; Lebrón-Aguilar, R.; Urpí-Sardà, M.; Llorach, R.; Andrés-Lacueva, C. Almond [Prunus dulcis (Mill.) D.A. Webb] polyphenols: from chemical characterization to targeted analysis of phenolic metabolites in humans. *Arch Biochem Biophys*. 2010;501(1):124-133.
33. Garrido, I.; Urpí-Sardà, M.; Monagas, M.; Gómez-Cordovés, C.; Martín-Alvarez, P.J.; Llorach, R.; Bartolomé, B.; Andrés-Lacueva, C. Targeted analysis of conjugated and microbial-derived phenolic metabolites in human urine after consumption of an almond skin phenolic extract. *J Nutr*. 2010; 140(10):1799-1807.
34. Vallverdú-Queralt, A.; Jáuregui, O.; Medina-Remón, A.; Andrés-Lacueva, C.; Lamuela-Raventós, R.M. Improved characterization of tomato polyphenols using liquid chromatography/electrospray ionization linear ion trap quadrupole Orbitrap mass spectrometry and liquid chromatography/electrospray ionization tandem mass spectrometry. *Rapid Commun Mass Spectrom*. 2010;24(20):2986-2992.
35. Llorach, R.; Garrido, I.; Monagas, M.; Urpí-Sardà, M.; Tulipani, S.; Bartolome, B.; Andres-Lacueva C. Metabolomics study of human urinary metabolome modifications after intake of almond [Prunus dulcis (Mill.) DA Webb] skin polyphenols. *J Proteome Res*. 2010;9(11):5859-5867.
36. Medina-Remón, A.; Zamora-Ros, R.; Rotchés-Ribalta, M.; Andres-Lacueva, C.; Martínez-González, M.A.; Covas, M.I.; Corella, D.; Salas-Salvadó, J.; Gómez-Gracia, E.; Ruiz-Gutiérrez, V.; García de la Corte, F.J.; Fiol, M.; Pena, M.A.; Saez, G.T.; Ros, E.; Serra-Majem, L.; Pinto, X., Warnberg, J.; Estruch, R.; Lamuela-Raventos, R.M.; on behalf of the PREDIMED Study Investigators. Total polyphenol excretion and blood pressure in subjects at high cardiovascular risk. *Nutr Metab Cardiovasc Dis*. 2010 [in press].
37. Visioli, F.; Alarcón de la Lastra, C.; Andres-Lacueva, C.; Aviram, M.; Calhau, C.; Cassano, A.; d'Archivio, M.; Faria, A.; Favé, G.; Fogliano, V.; Llorach, R.; Vitaglione, P.; Zoratti, M.; Edeas, M. Polyphenols and human health: a prospectus. *Crit Rev Food Sci Nutr*. 2010 [in press].
38. Estruch, R.; Sacanella, E.; Mota, F.; Chiva-Blanch, G.; Antunez, E.; Casals, E.; Deulofeu, R.; Rotilio, D.; Andrés-Lacueva, C.; Lamuela-Raventós, RM.; De Gaetano, G.; Urbano-Márquez, A. Moderate consumption of red wine, but not gin, decreases erythrocyte superoxide dismutase activity: A randomised cross-over trial. *Nutr Metab Cardiovasc Dis*. 2011; 21(1):46-53.

39. Chiva-Blanch, G.; Urpí-Sardà, M.; Rotchés-Ribalta, M.; Zamora-Ros, R.; Llorach, R.; Lamuela-Raventós, R.M.; Estruch, R.; Andrés-Lacueva, C. Determination of resveratrol and piceid in beer matrices by solid-phase extraction and liquid chromatography-tandem mass spectrometry. *J Chromatogr A*. 2011; 1218(5):698-705.
40. Vallverdú-Queralt, A.; Medina-Remón, A.; Andres-Lacueva, C.; Lamuela-Raventos, R.M. Changes in phenolic profile and antioxidant activity during production of diced tomatoes. *Food Chemistry*. 2011;126(4): 1700 – 1707.
41. Martinez, N.; Urpi-Sarda, M.; Martinez-Gonzalez, M.A.; Andres-Lacueva, C.; Mitjavila, M.T. Dealcoholised beers reduce atherosclerosis and expression of adhesion molecules in apoE-deficient mice. *Bf J Nutr*. 2011;105(5):721-730.
42. Khan, N.; Monagas, M.; Andrés-Lacueva, C.; Casas, R.; Urpí-Sardà, M.; Lamuela-Raventós, R.M.; Estruch, R. Regular consumption of cocoa powder with milk increases HDL cholesterol and reduces oxidized LDL levels in subjects at high-risk of cardiovascular disease. *Nutr Metab Cardiovasc Dis*. 2011 [in press].

In: Biochemistry Researcher Biographical Sketches ... ISBN: 978-1-62081-308-9
Editors: J. T. Walker and C. M. Jones

Chapter 2

Gry Aletta Bjorlykke

Affiliation:

Institute of Marine Research

Address:

P. 0. Box 1870 N-5817 Bergen, Norway

Date of Birth:

13.06.84

Education:

Bachelor and Master in Food Science at Norwegian University of Life Sciences

Publications Last 3 Years:

Bjorlykke, G. A., Roth, B., Seirheim, 0., Kvamme, B. 0., Slinde, E., 2011, The effects of carbon monoxide on Atlantic salmon (Salmo salar L.), *Food chemistry*, volume 127, issue 4, pp 1706-1711.

Volden, J., Bjorlykke, G.A., Sorheim, 0., Egelandsdal, B., Slinde, E. 2009. Pro-oxidative properties of monocarboxy myoglobin in systems with added polyunsaturated fatty acids. *Proc. International Congress of meat science and technology*, 16-21 August 2009, PE4.69, Copenhagen, volume 55 .

In: Biochemistry Researcher Biographical Sketches ... ISBN: 978-1-62081-308-9
Editors: J. T. Walker and C. M. Jones

Chapter 3

Victorio Cadierno

Affiliation:

Departamento de Química Orgánica e Inorgánica. Universidad de Oviedo

Address:

Departamento de Química Orgánica e Inorgánica. Facultad de Química. Universidad de Oviedo. Julian Claveria 8, E-33006, Oviedo, Spain

Date of Birth:

16/03/1969

Education:

Ph. D. in Chemistry (1996)

Research and Professional Experience:

Organometallic Chemistry, Homogeneous Catalysis, Green Chemistry

Professional Appointments:

Associate Professor of Inorganic Chemistry. Universidad de Oviedo (Spain)

Honors:

Spanish Royal Society of Chemistry (RSEQ) Young Investigator Award (2002).

Publications Last 3 Years:

"Allenylidene and higher cumulenylidene complexes". V. Cadierno, J. Gimeno. *Chem. Rev.*, 2009, 109, 3512-3560.

"Microwave-assisted InCl3-catalyzed Meyer-Schuster rearrangement of propargylic aryl carbinols in aqueous media: A green approach to a,3-unsaturated carbonyl compounds". V. Cadierno, J. Francos, J. Gimeno. *Tetrahedron Lett.*, 2009, *50*, 4773-4776.

"Alkyne cyclotrimerizations in environmentally benign media: Current state of the art". V. Cadierno, P. Crochet, S.E. García-Garrido. In Green Chemistry Research Trends; J.T. Pearlman Ed.; Nova Science Publishers, New York: USA; 2009, Chapter 3, 97-130.

"Ruthenium-catalyzed redox isomerization/transfer hydrogenation in organic and aqueous media: A one-pot tandem process for the reduction of allylic alcohols". V. Cadierno, P. Crochet, J. Francos, S.E. García-Garrido, J. Gimeno, N. Nebra. *Green Chem.*, 2009, *11*, 1992-2000.

"Ruthenium-catalyzed synthesis of 3-oxo esters in aqueous medium: Scope and limitations". V. Cadierno, J. Francos, J. Gimeno. *Green Chem.*, 2010, *12*, 135-143. Featured as a "Hot Article".

"Access to the first (iminophosphoranyl)(selenophosphoranyl)methane ligands Ph2P(=Se)CH2P(=NR)Ph2: Coordination of their methanide and methandiide anions to ruthenium". V. Cadierno, J. Díez, J. García-Álvarez, J. Gimeno. *Dalton Trans.*, 2010, *39*, 941-956.

"One-pot three-component synthesis of tetrasubstituted N-H pyrroles from secondary propargylic alcohols, 1,3-dicarbonyl compounds and *tert*-butyl carbamate". V. Cadierno, J. Gimeno, N. Nebra. *J. Heterocycl. Chem.*, 2010, *47*, 233-236.

"Atom-economic transformations of propargylic alcohols catalyzed by the 16-electron allyl-ruthenium(II) complex [Ru(i^3-2-C3H4Me)(CO)(dppf)][SbF6] (dppf = 1,1´-bis(diphenylphosphino)ferrocene)". V. Cadierno, S.E. García-Garrido, J. Gimeno, N. Nebra. *Inorg. Chim. Acta*, 2010, *363*, 1912-1934.

"One-pot synthesis of bicyclic fused cyclopentenone derivatives from 1- ethynyl-cycloalkanols and aldehydes". V. Cadierno, J. Gimeno, N. Nebra. *ChemCatChem*, 2010, *2*, 519-522.

"Metal-catalyzed transformations of propargylic alcohols into a,3-unsaturated carbonyl compounds: From the Meyer-Schuster and Rupe rearrangements to redox isomerization". V. Cadierno, P. Crochet, S.E. García-Garrido, J. Gimeno. *Dalton Trans.*, 2010, *39*, 4015-4031.

"Novel push-pull butadienes derived from 1,1-diaryl-2-propyn-1-ols and 1,1,1,5,5,5-hexafluoro-2,4-pentanedione: Synthesis, absorption spectra and solvatochromic behaviour". J. Borge, V. Cadierno, J. Díez. S.E. García-Garrido, J. Gimeno. *Dyes Pigm.*, 2010, *87*, 209-217.

"Ruthenium/TFA-catalyzed regioselective C-3-alkylation of indoles with terminal alkynes in water: Efficient and unprecedented access to 3-(1-methylalkyl)-1*H*-indoles". V. Cadierno, J. Francos, J. Gimeno. *Chem. Commun.*, 2010, *46*, 4175-4177.

"Bis(allyl)-ruthenium(IV) complexes containing water-soluble phosphane ligands: Synthesis, structure and application as catalysts in the selective hydration of organonitriles into amides". V. Cadierno, J. Diez, J. Francos, J. Gimeno. *Chem. Eur. J.*, 2010, *16*, 9808-9817.

"Arene-ruthenium(II) complexes containing amino-phosphine ligands as catalysts for nitrile hydration reactions". R. Garcia-Álvarez, J. Diez, P. Crochet, V. Cadierno. *Organometallics*, 2010, *29*, 3955-3965.

"All-carbon-substituted allenylidene and related cumulenylidene ligands". V. Cadierno, S.E. Garcia-Garrido. In Transition Metal Complexes of Neutral η1-Carbon Ligands; R. Chauvin & Y. Canac Eds.; Springer-Verlag, Berlin/Heidelberg: Germany; 2010, 151- 218 (Top. Organomet. Chem. vol. 30).

"Heteroatom-conjugated allenylidene and related cumulenylidene ligands". V. Cadierno, S.E. Garcia-Garrido. In Transition Metal Complexes of Neutral η1-Carbon Ligands; R. Chauvin & Y. Canac Eds.; Springer-Verlag, Berlin/Heidelberg: Germany; 2010, 219-252 (Top. Organomet. Chem. vol. 30).

"Metal-catalyzed reactions in water under MW irradiation". V. Cadierno, P. Crochet, S.E. Garcia-Garrido. In Aqueous Microwave Assisted Chemistry: Synthesis and Catalysis; V. Polshettiwar & R.S. Varma Eds.; RSC Publishing, Cambridge: UK; 2010, Chapter 2, 10-54.

"Palladium-catalyzed cycloisomerization of (*Z*)-enynols into furans using green solvents: Glycerol *vs.* water". J. Francos, V. Cadierno. *Green Chem.*, 2010, *12*, 1552- 1555.

"Iridium(I)-catalyzed coupling of (*Z*)-2-en-4-yn-1-ols with activated alkynes: A new synthetic route of 7-oxanorbornadienes". A.E. Diaz-Álvarez, P. Crochet, V. Cadierno. *Adv. Synth. Catal.*, 2010, *352*, 2427-2431.

"Cationic rhodium(I)-diolefin complexes containing an optically active *C*2-symmetric bis(sulfoximine) ligand: Synthesis and catalytic activity". V. Cadierno, J. Diez, S.E. Garcia-Garrido, J. Gimeno, A. Pizzano. *Polyhedron*, 2010, *29*, 3380-3386.

"Ruthenium-catalyzed intermolecular [2 + 2 + 2] alkyne cyclotrimerization in aqueous media under microwave-irradiation". V. Cadierno, J. Francos, S.E. Garcia-Garrido, J. Gimeno. *Green Chem. Lett. Rev.*, 2011, *4*, 55-61.

"Chemistry by nanocatalysis: First example of a solid-supported RAPTA complex for organic reactions in aqueous medium". S.E. Garcia-Garrido, J. Francos, V. Cadierno, J.M. Basset, V. Polshettiwar. *ChemSusChem*, 2011, *4*, 104-111.

"Ruthenium(IV) catalysts for the selective estragole to *trans*-anethole isomerization in environmentally friendly media". B. Lastra-Barreira, J. Francos, P. Crochet, V. Cadierno. *Green Chem.*, 2011, *13*, 307-313.

"Ruthenium(IV)-catalyzed Markovnikov addition of carboxylic acids to terminal alkynes in aqueous medium". V. Cadierno, J. Francos, J. Gimeno. *Organometallics*, 2011, *30*, 852-862.

"Access to unusual polycyclic spiro enones from 2,2′-bis(allyloxy)-1,1′-binaphthyls using Grubbs′ catalysts: An unprecedented one-pot RCM/Claisen sequence". E. Piedra, J. Francos, N. Nebra, F.J. Suárez, J. Díez, V. Cadierno. *Chem. Commun.*, 2011, *47*, 7866-7868.

"Ruthenium(IV)-catalyzed isomerization of the C=C bond of *O*-allylic substrates: A theoretical and experimental study". A. Varela-Álvarez, J.A. Sordo, E. Piedra, N. Nebra, V. Cadierno, J. Gimeno. *Chem. Eur. J.*, 2011, *17*, 10583-10599.

"Ibuprofenamide: a convenient method of synthesis by catalytic hydration of 2-(4-isobutylphenyl)propionitrile in pure aqueous medium". R. García-Álvarez, J. Francos, P. Crochet, V. Cadierno. *Tetrahedron Lett.*, 2011, *52*, 4218-4220.

"Ruthenium-catalyzed reduction of allylic alcohols using glycerol as solvent and hydrogen donor". A.E. Díaz-Álvarez, P. Crochet, V. Cadierno. *Catal. Commun.*, 2011, *13*, 91-96.

"Conjugate addition of arylboronic acids to *α,β*-unsaturated carbonyl compounds in aqueous medium using Pd(II) complexes with dihydroxy-2,2´-bipyridine ligands: Homogeneous or heterogeneous nano-catalysis?". E. Tomás-Mendivil, J. Díez, V. Cadierno. *Catal. Sci. Technol.*, in press. (Invited manuscript for the special issue: Heterogeneous Catalysis for Fine Chemicals)

"Expeditious entry to novel 2-methylene-2,3-dihydrofuro[3,2-*c*]chromen-2-ones from 6-chloro-4-hydroxychromen-2-one and propargylic alcohols". N. Nebra, A.E. DíazÁlvarez, J. Díez, V. Cadierno. *Molecules*, 2011, *16*, 6470-6480. (Invited manuscript for the special issue: Heterocycles)

"Ruthenium complexes containing iminophosphorane ligands: Synthesis, structure, reactivity and applications". V. Cadierno, J. García-Álvarez. In *Ruthenium: Properties, Production and Applications*; D.B. Watson Ed.; Nova Science Publishers, New York: USA; 2011, 189-220.

"Glycerol and derived solvents: New sustainable reaction media for organic synthesis". A.E. Díaz-Álvarez, J. Francos, B. Lastra-Barreira, P. Crochet, V. Cadierno. *Chem. Commun.*, 2011, *47*, 6208-6227.

"MW-assisted Meyer-Schuster rearrangement of propargylic alcohols catalyzed by the oxovanadate complex [V(O)Cl(OEt)2]". A. Antiñolo, F. Carrillo-Hermosilla, V. Cadierno, J. García-Álvarez, A. Otero. *ChemCatChem*, in press.

"Arene-ruthenium(II) complexes containing inexpensive tris(dimethylamino)phosphine: Highly efficient catalysts for the selective hydration of nitriles into amides". R. García-Álvarez, J. Díez, P. Crochet, V. Cadierno. *Organometallics*, accepted.

"A general route for the stereoselective synthesis of (*E*)-(1-propenyl)phenyl esters by catalytic C=C bond isomerization". A.E. Díaz-Álvarez, P. Crochet, V. Cadierno. *ChemCatChem*, submitted.

In: Biochemistry Researcher Biographical Sketches ... ISBN: 978-1-62081-308-9
Editors: J. T. Walker and C. M. Jones

Chapter 4

Wisit Cheungpasitporn

Affiliation:

Department of Internal Medicine, Bassett Medical Center and Columbia University College of Physicians and Surgeons, Cooperstown, New York13326, USA

Address:

10 Fernleigh Dr. APT#C3. Cooperstown, NY, 13326 USA

Date of Birth:

07/15/1983

Education:

Doctor of Medicine and Internal Medicine

Research and Professional Experience:

1. Clinical and Basic Reseacher, July 2010- present Bassett Medical Center and *Columbia University College of Physicians and Surgeons*, Cooperstown, New York
2. Editor for Emergency Medicine journalm2011- present
3. Reviewer- *Annals of Internal Medicine* 2010-2011, -The *Journal of General Internal Medicine, 2010-2011; -British Journal of Medical Practitioners,* 2010-2011.
4. Clinical Researcher July-September2009 Allergy and Clinical Immunology Unit. King Chulalongkorn Memorial Hospital, Bangkok, Thailand

Professional Appointments:

1. Professional and Society Memberships, Member, Medical Council of Thailand, 2007-present
   "National Public Health Officer" of IFMSA-Thailand, 2004-2005
   (International Federation of Medical Students' Association)
2. Editor for Emergency Medicine journal, 2011- present
3. Reviewer, *Annals of Internal Medicine*, 2010-2011
   -The *Journal of General Internal Medicine, 2010-2011*
   -British Journal of Medical Practitioners, 2010-2011

Honors:

"Highest Award In Emergency Medicine", m2007
Faculty of Medicine, Chulalongkorn University, *Bangkok, Thailand*
"Outstanding Performance In Pediatrics Clerkship", 2007
Faculty of Pediatrics, Chulalongkorn University, *Bangkok, Thailand*

Publications Last 3 Years:

Suksaranjit P, Cheungpasitporn W, Bischof EF, Marx H, Venkatachalam S, Rimmerman CM. The negative U wave in the setting of demand ischemia (AUGUST 2011). *Cleve Clin J Med.* 2011 Dec;78(12):785.

Ungprasert P, Ratanapo S, Cheungpasitporn W. Management in Thoracic Aorta Mural Thrombi: Evidence Based Medicine and Controversy. *Emergency Medicine.* 2011; 1(1):1000e104.

Chanprasert S, Cheungpasitporn W, Eldred AK. Coexistent Membranous Nephropathy with Doubly ANCA-Associated Crescentic Glomerulonephritis: A Case Report and Review of Literature. *J Nephrol Therapeutic* 2011, 1:106. 2011; 1(2).

Cheungpasitporn W, Suksaranjit P, Chanprasert S. Hypercalcemia in human immunodeficiency virus-related lymphoma and valacyclovir toxicity. *Am J Med Sci.* 2011 Dec;342(6):539.

Cheungpasitporn W, Suksaranjit P, Chanprasert S. Acute kidney injury from bilateral ureteral calcium stones in the setting of primary hyperparathyroidism. *Am J Emerg Med.* 2011 Nov 17. [Epub ahead of print]

Cheungpasitporn W, Kaewpoowat Q, Prasidthrathsint K, Suksaranjit P. Warfarin-related nephropathy can be mimicked by an interaction between sulfonamide derivatives and vitamin K antagonists. *Kidney Int.* 2011 Dec;80(11):1246.

Cheungpasitporn W, Alvi M, Medarametla V. Association between acute urinary retention and Guillain-Barré syndrome. *Am J Emerg Med.* 2011 Oct 25. [Epub ahead of print]

Cheungpasitporn W, Kaewpoowat Q, Suksaranjit P, Medarametla V, Chikoti S, Bischof EF. Pulmonary Arteriovenous Malformation in A Woman With Severe Mitral Valve Stenosis. ConsultantLive. 2011.

Cheungpasitporn W, Jirajariyavej T, Howarth CB, Rosen RM. Henoch-Schönlein purpura in an older man presenting as rectal bleeding and IgA mesangioproliferative glomerulonephritis: a case report. *J Med Case Reports*. 2011 Aug 10;5:364.

Cheungpasitporn W, Horne JM, Howarth CB. Adrenocortical carcinoma presenting as varicocele and renal vein thrombosis: a case report. *J Med Case Reports*. 2011 Aug 1;5:337.

Cheungpasitporn W, Jirajariyavej T, Chanprasert S. Rate control medications for atrial fibrillation in the setting of hypercalcemia. *Am J Emerg Med*. 2011 Sep;29(7):830.

Techapornroong M, Akrawinthawong K, Cheungpasitporn W, Ruxrungtham K. Anaphylaxis: a ten years inpatient retrospective study. *Asian Pac J Allergy Immunol*. 2010 Dec;28(4):262-9.

Chantaphakul H., Cheungpasitporn W, Ruxrungtham K., Klaewsongkram J. Adverse Reaction to Anti-tuberculosis Drugs. *The Journal of Allergy and Clinical Immunology*. 2010; 125(2):Page AB154.

In: Biochemistry Researcher Biographical Sketches ... ISBN: 978-1-62081-308-9
Editors: J. T. Walker and C. M. Jones

Chapter 5

Marcio Jose da Silva

Affiliation:

Federal University of Viçosa

Address:

Av P H Rolfs Viçosa Minas Gerais

Date of Birth:

21/03/1965

Education:

post-doctor/2005-2006/Federal University of Minas Gerais
D.S 1999-2004//Federal University of Minas Gerais
M.S 1997-1999//Federal University of Minas Gerais

Research and Professional Experience:

Professor 2006-nowadays/ Federal University of Viçosa

Actuation fields: Teaching, Researches in biofuels production technologies; glycerol chemistry/olefin oxidation catalytic processes

Publications Last 3 Years:

1. Fernandes, Sergio Antonio; Cardoso, Abiney Lemos; da Silva, Márcio José; SILVA, M. J. A novel kinetic study of H3PW12O40 - catalyzed oleic acid esterification with methanol via 1H NMR spectroscopy. *Fuel Processing Technology JCR*,, v. 96, p. 98-103, 2012.
2. Fernandes, S.A.; SILVA, M. J.; NATALINO, R.; Jham G N ; Gazolla P A R . p-Sulfonic acid calix[n]arenes as homogeneous and recyclable organocatalysts for esterification reactions. *Tetrahedron Letters JCR*, v. xx, p. xx-xx, 2012.
3. Oliveira, LCA ; Coelho J ; Sousa P ; Silva C ; SILVA, M. J. . *Beta-pinene oxidation by hydrogen peroxide catalyzed by modified niobium -MCM* (in press). Applied Catalysis. A, General (Print) JCR,, v. xx, p. xx-xx, 2012.
4. SILVA, M. J.; Lair, L.O.; Gonçalves, C.E.M.. Novel esterification of glycerol catalysed by tin chloride (II): a recyclable and less corrosive process for production of bio-additives. *Catalysis Letters JCR*, v. 141, p. 1111-1117, 2011.
5. SILVA, M. J.; Cardoso, Abiney Lemos; Figueiredo, A. P.; SILVA, M. L. ; NATALINO, R.. Effect of water on the ethanolysis of waste cooking soybean oil using a Tin(II) chloride catalyst. *Journal of the American Oil Chemists' Society JCR*, v. 88, p. 1431-1437, 2011.
6. Carari, Danieli M.; da Silva, Márcio J.; SILVA, M. J. A Highly Selective Pd(OAc)2/Pyridine/K2CO3 System for Oxidation of Terpenic Alcohols by Dioxygen. *Catalysis Letters JCR*, v. xx, p. xx-xx, 2011.
7. SILVA, M. J; Cardoso, Abiney Lemos; NATALINO, R. Tin Catalysed Esterification of Free Fatty Acids. *International Journal of Chemical Reactor Engineering JCR*, v. 8, p. 1-12, 2010.
8. Godói Silva, Vinícius Wilker ; Laier, Letícia Oliveira ; Silva, Márcio José da ; SILVA, M. J.. Novel H3PW12O40: Catalysed Esterification Reactions of Fatty Acids at Room Temperature for Biodiesel *Production. Catalysis Letters JCR*, v. 135, p. 207-211, 2010.
9. SILVA, M. J. ; Cardoso, Abiney Lemos ; Neves, S. C. G. . Kinetic Study of Alcoholysis of the Fatty Acids Catalyzed by Tin Chloride (II): An Alternative Catalyst for Biodiesel Production. *Energy & Fuels (Print) JCR*, v. 23, p. 1718/10.1021/ef-1722, 2009.
10. SILVA, M. J.; Carari, D. M.; Teixeira, R. R. Pd(OAc)2/M(NO3)n (M = Cu(II), Fe(III); n = 2, 3): Kinetic investigations of an alternative Wacker system for the oxidation of natural olefins. *Journal of Organometallic Chemistry (Print) JCR*, v. 694, p. 3254-3261, 2009.
11. SILVA, M. J.; LAGO, R. M.; GROSSI, C.; ARAUJO, M. H. Sulfonated polystyrene: A catalyst with acid and superabsorbent properties for the esterification of fatty acids. Fuel (Guildford) JCR, v. 89, p. 257-259, 2009.
12. SILVA, M. J.; OLIVEIRA, A. A.; SILVA, M. L.. Palladium-Catalysed Oxidation of Bicycle Monoterpenes by Hydrogen Peroxide in Acetonitrile Solutions: A Metal Reoxidant-Free and Environmentally Benign Oxidative Process. *Catalysis Letters JCR*, v. 130, p. 424-431, 2009.
13. SILVA, M. J.; AUGUSTI, R.; Cardoso, Abiney Lemos. Investigation on the esterification of fatty acids catalyzed by the H3PW12O40 heteropolyacid. *Journal of the American Oil Chemists' Society JCR*, v. 85, p. 555-560, 2008.

14. Cardoso, Abiney L.; Neves, Soraia Cristina Gonzaga ; da Silva, Marcio J. ; SILVA, M. J. Esterification of Oleic Acid for Biodiesel Production Catalyzed by SnCl2: A *Kinetic Investigation. Energies* (Basel) JCR, v. 1, p. 79-92, 2008.

In: Biochemistry Researcher Biographical Sketches ... ISBN: 978-1-62081-308-9
Editors: J. T. Walker and C. M. Jones

Chapter 6

Isabelle Denis

Affiliation:

Unité de Nutrition et régulations lipidiques des fonctions cérébrales (NuReLiCe), INRA. 78352 Jouy-en-Josas cedex. France.

Address:

Tel : 33 1 34 65 23 13
Fax: 33 1 34 65 23 11
Isabelle.denis@jouy.inra.fr

Date of Birth:

02.14.1964

Education:

PhD, HDR (habilitation à diriger des recherches) in medicine and biology (Université de Versailles), PhD in physiology (Université Paris XI), Master in Nutrition (Université Paris XI).

Research and Professional Experience:

Researcher (Chargée de Recherche 1) at INRA (National Institute of agronomical research), France (from 1990).

Professional Appointments:

Nutritional influence of n-3 PUFA on brain function (from 2000).

Publications Last 3 Years:

M.Lavialle, I.Denis, P.Guesnet, S.Vancassel. 2010. Involvement of omega-3 fatty acids in emotional responses and hyperactive symptoms. review *J Nut Biochem 21, 899-905.*

Potier B, Billard J-M, Rivière S, Sinet P-M, Denis, I., Champeil-Potokar G., Grintal B., Jouvenceau A., Kollen M., Dutar P. 2010. Reduction of glutamate uptake is associated with extrasynaptic NMDA and metabotropic glutamate receptor activation at the hippocampal CA1 synapse of aged rats. *Aging Cell 9, 722-735.*

Mathieu G, Denis S, Langelier B, Denis I, Lavialle M and Vancassel S. 2010. DHA enhances the noradrenaline release in SH-SY5Y cells. *Neurochem Int. 56, 94-100.*

Grintal B, Champeil-Potokar G, Lavialle M, Vancassel S, Breton S, Denis I. 2009. Inhibition of astroglial glutamate transport by polyunsaturated fatty acids: evidence for a signalling role of docosahexaenoic acid. *Neurochem. Int.54: 535-543*

Lavialle M, Champeil G, Alessandri JM, Balasse L, Guesnet P, Linard A, Lallemand MS, Papillon C, Pévet P, Vancassel S, Vivien-Roels B, Denis I. 2008. N-3 fatty acid deficient diet affects daily melatonin rhythm in Syrian hamster: incidences on hyperactivation and photic entrainment of locomotor activity. *J.Nutr.138 : 1719-1724.*

F. Pifferi, M. Jouin, JM. Alessandri, U. Haedke, F. Roux, N. Perrière, I. Denis, M. Lavialle' and P. Guesnet. 2007. N-3 fatty acids modulate brain glucose transport in endothelial cells of the blood-brain barrier. *Prostag Leukotr Ess Fatty Acids,* 77, 279-286.

Champeil-Potokar G, Chaumontet C, Guesnet P, Lavialle M, Denis I. 2006. Docosahexaenoic acid (DHA, 22:6n-3) enrichment of membrane phospholipids increases gap junction coupling capacity in cultured astrocytes. *Eur.J.Neurosci,* 24 : 3084-3090.

In: Biochemistry Researcher Biographical Sketches ... ISBN: 978-1-62081-308-9
Editors: J. T. Walker and C. M. Jones

Chapter 7

Mark D. Distefano

Address:

Office: 207 Pleasant Street SE, Minneapolis, MN 55455
Home: 4820 Narcissus Ct. N., Plymouth, MN 55446
Office: 612-624-0544
Email: diste001@umn.edu

Education:

Ph. D. in Chemistry (June, 1989)
Massachusetts Institute of Technology
B.A., Double Major in Chemistry and Biochemistry (June, 1984)
University of California, Berkeley

Research and Professional Experience:

2009 – present Professor, Department of Chemistry, University of Minnesota, Minneapolis, MN
1998 – 2009 Associate Professor, Department of Chemistry, University of Minnesota, Minneapolis, MN
1992 – 1998 Assistant Professor, Department of Chemistry, University of Minnesota, Minneapolis, MN
1989 – 1992 Postdoctoral Fellow, Department of Chemistry, California Institute of Technology, Pasadena, CA
1987 – 1989 Visiting Scientist, Department of Biological Chemistry and Molecular Pharmacology, Harvard Medical School, Boston, MA Other appointments
2006 – present Medicinal Chemistry Graduate Program Member
2005 – present Microbial Biotechnology Graduate Program Member
2005 – present University of Minnesota Biotechnology Institute Member

2003 – 2006 Director of Graduate Studies in Chemistry
2001 Visiting Scientist, Genentech Corporation, South San Francisco, CA
1995 – present University of Minnesota Cancer Center Member

Selected awards

2011 Distinguished McKnight University Professor
2011 Merck Professor of Chemistry
2006 Horace T. Morse-University of Minnesota Distinguished Teaching, Professor
2004 George W. Taylor/ITAS Award for Distinguished Teaching
1994 National Science Foundation CAREER Award
1994 American Cancer Society Junior Faculty Research Award
1992 3M Alumni Assistant Professor
1989 Damon Runyon - Walter Winchell Cancer Foundation Fellowship

Professional organizations

American Chemical Society
American Associate for the Advancement of Science
Professional Appointments:

Publications Last 3 Years:

1. Biochemical probes

1. Wollack, J. W.; Zeliadt, N. A.; Mullen, D. G.; Amundson, G.; Geier, S.; Falkum, S.; Wattenberg, E. V.; Barany, G.; Distefano, M. D., Multifunctional prenylated peptides for live cell analysis. *Journal of the American Chemical Society* 2009, 131, 7293–7303. DOI: 10.1021/ja805174z
2. DeGraw, A. J.; Zhao, Z.; Hsieh, J.; Jefferies, M.; Distefano, M. D.; Strickland, C. L.; Shintani, D.; Nural, H.; McMahan, C.; Xie, W., A photoactive isoprenoid diphosphate analogue containing a stable phosphonate linkage: synthesis and structural biochemical studies with prenyltransferases. *Journal of Organic Chemistry* 2007, 72, (13), 4587 - 4595. DOI: 10.1021/jo0623033 (Cover feature).

Biotechnology

Duckworth, B. P.; Chen, Y.; Sham, Y.; Mueller, J. D.; Taton, T. A.; Distefano, M. D., A Universal Method for the Preparation of Covalent Protein-DNA Conjugates for Use in Creating Protein Nanostructures. *Angewante Chemie (Int. Ed)* 2007, 46, 8819-8822. DOI: 10.1002/anie.200701942

Duckworth, B. P.; Xu, J.; Taton, T. A.; Guo, A.; Distefano, M. D., Site-Specific, Covalent Attachment of Proteins to a Solid Surface. *Bioconjugate Chemistry* 2006, 17, (4), 967-974. DOI: 10.1021/bc060125e

Enzymology

Lenevich, S.; Hosokawa, A.; Cramer, C. J.; Distefano, M. D., Transition State Analysis of Model and Enzymatic Prenylation Reactions. *Journal of the American Chemical Society* 2007, 129, 5796-5797. DOI: 10.1021/ja069119j

Turek-Etienne, T. C.; Strickland, C. L.; Distefano, M. D., Biochemical and Structural Studies with Prenyl Diphosphate Analogues Provide Insights into Isoprenoid Recognition by Protein Farnesyl Transferase. *Biochemistry* 2003, 42, (13), 3716-3724. DOI: 10.1021/bi0266838

Clausen, V. A.; Edelstein, R. L.; Distefano, M. D., Stereochemical Analysis of the Reaction Catalyzed by Human Protein Geranylgeranyl Transferase. *Biochemistry* 2001, 40, (13), 3920-3930. DOI: 10.1021/bi002011a

Protein engineering

Qi, D.; Tann, C.-M.; Häring, D.; Distefano, M. D., Generation of new enzymes via covalent modification of existing proteins. *Chemical Reviews* 2001, 101, (10), 3081-3111. DOI: 10.1021/cr000059o

Häring, D.; Distefano, M. D., Enzymes by Design: Chemogenetic Assembly of Transamination Active Sites Containing Lysine Residues for Covalent Catalysis. *Bioconjugate Chemistry* 2001, 12, (3), 385-390. DOI: 10.1021/bc000117c

Davies, R. R.; Distefano, M. D., A Semisynthetic Metalloenzyme Based on a Protein Cavity That Catalyzes the Enantioselective Hydrolysis of Ester and Amide Substrates. *Journal of the American Chemical Society* 1997, 119, (48), 11643-11652. DOI: 10.1021/ja970820k

Complete list (in reverse chronological order)

102. Kyro, K.; Manandhar, S. P.; Mullen, D.; Schmidt, W. K.; Distefano, M. D., Photoaffinity Labeling of Ras Converting Enzyme using Peptide Substrates that Incorporate Benzoylphenylalanine (Bpa) Residues: Improved Labeling and Structural Implications. *Bioorganic and Medicinal Chemistry* 2011, in press.

101. Ochocki, J. D.; Igbavboa, U.; Wood, W. G.; Wattenberg, E. V.; Distefano, M. D., Evaluation of prenylated peptides for use in cellular imaging and biochemical analysis. *Methods in Molecular Biology* 2011, in press.

100. Ochocki, J. D.; Mullen, D. G.; Wattenberg, E. V.; Distefano, M. D., Evaluation of a cell penetrating prenylated peptide lacking an intrinsic fluorophore via in situ click reaction. *Bioorganic and Medicinal Chemistry Letters* 2011, *21*, 4998-5001, DOI: 10.1016/j.bmcl.2011.04.138.

99. Gaffarogullari, E. C.; Masterson, L. R.; Metcalfe, E. E.; Traaseth, N. J.; Balatri, E.; Musa, M. M.;Mullen, D.; Distefano, M. D.; Veglia, G., A Myristoyl/Phosphoserine Switch Controls cAMPDependent Protein Kinase Association to Membranes. *Journal of Molecular Biology* 2011, *411*, 823-836, DOI: 10.1016/j.jmb.2011.06.034.

98. Mullen, D.; Kyro, K.; Hauser, M.; Becker, J. M.; Naider, F.; Distefano, M. D., Synthesis of a-factor peptide from Saccharomyces cerevisiae and photoactive analogues via Fmoc solid phase methodology. *Bioorganic and Medicinal Chemistry* 2011, *19*, 490-497, DOI:10.1016/j.bmc.2010.11.006.
97. Lenevich, S.; Distefano, M. D., NMR-based quantification of organic diphosphates. *Analytical Biochemistry* 2011, *408*, 316-320, DOI: 10.1016/j.ab.2010.08.030.
96. Wang, Y.-C.; Distefano, M. Preparation and Analysis of One-Bead One-Compound Libraries for Studying Protein Prenylation Specificity. In *Building Bridges: Proceedings of the 22nd American Peptide Symposium*; Lebl, M., Ed.; Prompt Scientific Publishing: San Diego, 2011, p 276-277.
95. Rashidian, M.; Distefano, M. D. Site-Specific Protein Labeling Via Oxime and Hydrazone Ligations Using Protein Farnesyltransferases. In *Building Bridges: Proceedings of the 22nd American Peptide Symposium*; Lebl, M., Ed.; Prompt Scientific Publishing: San Diego, 2011, p 88-89.
94. Kyro, K.; Schmidt, W. K.; Distefano, M. D. Design and Application of Bpa-Containing Peptides for Studying Ras-Converting Enzyme. In *Building Bridges: Proceedings of the 22nd American Peptide Symposium*; Lebl, M., Ed.; Prompt Scientific Publishing: San Diego, 2011, p 268-269.
93. Díaz-Rodríguez, V.; Mullen, D.; Distefano, M. D. Synthesis of a-Factor and a-Factor Analogs Using a Cysteine-Anchoring Method. In *Building Bridges: Proceedings of the 22nd American Peptide Symposium*; Lebl, M., Ed.; Prompt Scientific Publishing: San Diego, 2011, p 92-93.
92. Abate-Pella, D.; Distefano, M. D. Photocaged Cysteine Peptides for Studying Protein Farnesylation. In *Building Bridges: Proceedings of the 22nd American Peptide Symposium*; Lebl, M., Ed.; Prompt Scientific Publishing: San Diego, 2011, p 90-91.
91. Rashidian, M.; Dozier, J. K.; Lenevich, S.; Distefano, M. D., Selective Labeling of Polypeptides Using Protein Farnesyltransferase via Rapid Oxime Ligation. *Chemical Communications* 2010, *46*, 8998-9000, DOI: 10.1039/C0CC03305G.
90. DeGraw, A. J.; Palsuledesai, C.; Ochocki, J. D.; Dozier, J. K.; Lenevich, S.; Rashidian, M.; Distefano, M. D., Evaluation of alkyne-modified isoprenoids as chemical reporters of protein prenylation. *Chem. Bio. Drug. Des.* 2010, *76*, 460-471, DOI: 10.1111/j.1747-0285.2010.01037.x.
89. Kyro, K.; Manandhar, S.; Mullen, D.; Schmidt, W. K.; Distefano, M. D., Photoaffinity Labeling of Ras Converting Enzyme 1 (RCE1) using a Benzophenone-Containing Peptide Substrate. *Bioorganic and Medicinal Chemistry* 2010, *18*, 5675-5684, DOI:10.1016/j.bmc.2010.06.024.
88. Ochocki, J. D.; Wattenberg, E. V.; Distefano, M. D., Enlarging the scope of cell penetrating prenylated peptides to include farnesylated "CAAX" box sequences and diverse cell types. *Chem. Biol. Drug Des.* 2010, *76*, 107-115, DOI: 10.1111/j.1747-0285.2010.00992.x.
87. Lopez-Gallego, F.; Agger, S. A.; Abate-Pella, D.; Distefano, M. D.; Schmidt-Dannert, C., Sesquiterpene synthases Cop4 and Cop6 from Coprinus cinereus: catalytic promiscuity and cyclization of farnesyl pyrophosphate geometric isomers. *Chembiochem.* 2010, *11*, 1093-1106, DOI: 10.1002/cbic.200900671.

86. Mullen, D.; Weigel, B.; Barany, G.; Distefano, M. D., On-resin conversion of Cys(Acm)-containing peptides to their corresponding Cys(Scm) congeners: scope and limitations. *Journal of Peptide Science* 2010, 16, 219-222. DOI: 10.1002/psc.1223
85. DeGraw, A. J.; Keiser, M. J.; Ochocki, J. D.; Shoichet, B. K.; Distefano, M. D., Prediction andevaluation of protein farnesyltransferase inhibition by commercial drugs *Journal of Medicinal Chemistry* 2010, 53, 2464–2471. DOI: 10.1021/jm901613f
84. Wollack, J. W.; Zeliadt, N. A.; Mullen, D. G.; Wattenberg, E. V.; Barany, G.; Distefano, M. D.,Investigation of a Minimal Sequence for Cell-Penetrating Prenylated Peptides. *Bioorganic and Medicinal Chemistry Letters* 2010, 20, 161-163. DOI:10.1016/j.bmcl.2009.11.026
83. Hovlid, M. L.; Edelstein, R. L.; Henry, O.; Ochocki, J.; Talbot, T.; Lopez-Gallego, F.; Schmidt-Dannert, C.; Distefano, M. D., Synthesis and Applications of DATFP-Containing Photoactive Analogues of Farnesyl Diphosphate Containing Modified Linkages for Enhanced Stability. *Chemical Biology and Drug Design* 2010, 75, 51-67. DOI: 10.1111/j.1747-0285.2009.00914.x
82. Wollack, J.; Silvermann, J.; Petzold, C.; Mougous, J.; Distefano, M. D., A Minimalist Substrate for Enzymatic Peptide and Protein Conjugation. *ChemBiochem* 2009, 10, 2934-2943. DOI: 10.1002/cbic.200900566
81. Abate-Pella, D.; Zeliadt, N.; Mullen, D.; McDonald, P.; Pathak, P.; Dore, T.; Mueller, J.;Wattenberg, E.; Distefano, M. D., A Photolabile Thiol Protecting Group for Cellular Studies. In *Breaking Away: Proceedings of the 21st American Peptide Symposium*, Lebl, M., Ed. Prompt Scientific Publishing: San Diego, 2009; pp 362-363.
80. Wollack, J. W.; Katzenmeyer, J.; Ochocki, J. D.; Zeliadt, N. A.; Mullen, D. G.; Arriaga, E. A.;Wattenberg, E. V.; Distefano, M. D., Utilization of Hydrophobic Modification to Promote Internalization of a CAAX Box-containing Peptide. In *Breaking Away: Proceedings of the 21st American Peptide Symposium*, Lebl, M., Ed. Prompt Scientific Publishing: San Diego, 2009; pp 318-320.
79. Kyro, K.; Mullen, D.; Manandhar, S.; Walter K. Schmidt; Distefano, M. D., Photoactive Isoprenoid-containing Peptides. In *Breaking Away: Proceedings of the 21st American Peptide Symposium*, Lebl, M., Ed. Prompt Scientific Publishing: San Diego, 2009; pp 304-305.
78. Wollack, J. W.; Zeliadt, N. A.; Mullen, D. G.; Amundson, G.; Geier, S.; Falkum, S.; Wattenberg, E. V.; Barany, G.; Distefano, M. D., Multifunctional prenylated peptides for in vivo analysis. *Journal of the American Chemical Society* 2009, 131, 7293–7303. DOI: 10.1021/ja805174z
77. Henry, O.; Lopez-Gallego, F.; Agger, S. A.; Schmidt-Dannert, C.; Sen, S.; Shintani, D.; Cornish, K.; Distefano, M. D., A Versatile Photoactivatable Probe Designed to Label the Diphosphate Binding Site of Farnesyl Diphosphate Utilizing Enzymes. *Bioorganic & Medicinal Chemistry* 2009, 17, 4797-4805. DOI:10.1016/j.bmc.2009.04.034.
76. Chung, J. A.; Wollack, J. W.; Okesli, A.; Hovlid, M. L.; Chen, Y.; Mueller, J. D.; Distefano, M. Taton, T. A., Purification of prenylated proteins by affinity chromatography on cyclodextrinmodified agarose. *Analytical Biochemistry* 2009, 386, 1-8. DOI:10.1016/j.ab.2008.09.007
75. Sarikonda, G.; Wang, H.; Puan, K.-J.; Liu, X.-h.; Lee, H. K.; Song, Y.; Distefano, M. D.; Oldfield, Prestwich, G. D.; Morita, C. T., Photoaffinity Antigens for Human gamma-delta T Cells. *Journal of Immunology*. 2008, 181, (11), 7738-7750.

74. DeGraw, A. J.; Hast, M. A.; Xu, J.; Mullen, D.; Beese, L. S.; Barany, G.; Distefano, M. D., Caged Protein Prenyltransferase Substrates: Tools for Understanding Protein Prenylation. *Chemical Biology and Drug Design,* 2008, 72, 171-181. DOI: 10.1111/j.1747-0285.2008.00698.x

73. Xie, W.; McMahan, C. M.; Distefano, M. D.; DeGraw;, A. J.; Cornish, K.; Whalen, M. C.; Shintani, D. K., Initiation of rubber biosynthesis: in vitro comparisons of benzophenone-modified diphosphate analogues in three rubber-producing species. *Phytochemistry*, 2008, 69, 2539-2545. DOI:10.1016/j.phytochem.2008.07.011.

72. Lenevich, S.; Hosokawa, A.; Cramer, C. J.; Distefano, M. D., Transition State Analysis of Model and Enzymatic Prenylation Reactions. *Journal of the American Chemical Society* 2007, 129, 5796-5797. DOI: 10.1021/ja069119j

71. Hosokawa, A.; Wollack, J. W.; Zhang, Z.; Chen, L.; Barany, G.; Distefano, M. D., Evaluation of an alkyne-containing analogue of farnesyl diphosphate as a dual substrate for proteinprenyltransferases. *International Journal of Peptide Research* 2007, 13, (1-2), 345-354. DOI: 10.1007/s10989-007-9090-3

70. Duckworth, B. P.; Zhang, Z.; Hosokawa, A.; Distefano, M. D., Selective Labeling of Proteins using Protein Farnesyltransferase. *ChemBioChem* 2007, 8, (1), 98-105. DOI: 10.1002/cbic.200600340.

69. Duckworth, B. P.; Chen, Y.; Sham, Y.; Mueller, J. D.; Taton, T. A.; Distefano, M. D., A UniversalMethod for the Preparation of Covalent Protein-DNA Conjugates for Use in Creating Protein Nanostructures. *Angewante Chemie (Int. Ed)* 2007, 46, 8819-8822. DOI: 10.1002/anie.200701942

68. DeGraw, A. J.; Zhao, Z.; Hsieh, J.; Jefferies, M.; Distefano, M. D.; Strickland, C. L.; Shintani, D.;Nural, H.; McMahan, C.; Xie, W., A photoactive isoprenoid diphosphate analogue containing a stable phosphonate linkage: synthesis and structural biochemical studies with prenyltransferases. *Journal of Organic Chemistry* 2007, 72, (13), 4587 - 4595. DOI: 10.1021/jo0623033.

67. Xu, J.; DeGraw, A. J.; Duckworth, B. P.; Lenevich, S.; Tann, C.-M.; Jenson, E. C.; Gruber, S. J.;Barany, G.; Distefano, M. D., Synthesis and Reactivity of 6,7-Dihydrogeranylazides: Reagents for Primary Azide Incorporation into Peptides and Subsequent Staudinger Ligation. *Chemical Biology and Drug Design* 2006, 68, 85-96. DOI: 10.1111/j.1747-0285.2006.00420.x

66. Duckworth, B. P.; Xu, J.; Taton, T. A.; Guo, A.; Distefano, M. D., Site-Specific, Covalent Attachment of Proteins to a Solid Surface. *Bioconjugate Chemistry* 2006, 17, (4), 967-974. DOI: 10.1021/bc060125e

65. Distefano, M. D.; Albers, L. N.; Xu, J.-H., Protein Prenylation. In *Encyclopedic Reference of Genomics and Proteomics in Molecular Medicine*, ed.; Ruckpaul, K.; Ganten, D., SpringerVerlag: Heidelberg, Germany, 2006; pp 349-354.

64. Duckworth, B.; Distefano, M. D., Protein-based artificial enzymes. In *Artificial Enzymes*, ed.; Breslow, R., Wiley-VCH Verlag: Weinheim, Germany, 2005; p 109-132.

63. Rose, M. W.; Rose, N. D.; Boggs, J.; Lenevich, S.; Xu, J.; Barany, G.; Distefano, M. D.,Evaluation of geranylazide and farnesylazide diphosphate for incorporation of prenylazides into a CAAX box-containing peptide using protein farnesyltransferase. *Journal of Peptide Research* 2005, 65, 529-537. DOI: 10.1111/j.1399-3011.2005.00261.x

62. Rose, M. W.; Xu, J.; Kale, T. A.; O'Doherty, G.; Barany, G.; Distefano, M. D., Enzymatic incorporation of orthogonally reactive prenylazide groups into peptides via protein farnesyltransferase: implications for selective protein labeling. *Biopolymers* 2005, 80, (2-3), 164- 171.

61. Adams, D. N.; Kao, E. Y.-C.; Hypolite, C. L.; Distefano, M. D.; Hu, W.-S.; Letourneau, P. C.,Growth cones turn and migrate up an immobilized gradient of the laminin IKVAV peptide. *Journal of Neurobiology* 2004, 62, (1), 134-147. DOI: 10.1002/neu.20075.

60. Franklin, M. C.; Kadkhodayan, S.; Ackerly, H.; Alexandru, D.; Distefano, M. D.; Elliott, L. O.; Flygare, J. A.; Mausisa, G.; Okawa, D. C.; Ong, D.; Vucic, D.; Deshayes, K.; Fairbrother, W. J., Structure and Function Analysis of Peptide Antagonists of Melanoma Inhibitor of Apoptosis (MLIAP). *Biochemistry* 2003, 42, (27), 8223-8231. DOI: 10.1021/bi034227t

59. Kale, T. A.; Distefano, M. D., Diazotrifluoropropionamido-Containing Prenylcysteines: Syntheses and Applications for Studying Isoprenoid-Protein Interactions. *Organic Letters* 2003, 5, (5), 609- 612. DOI: 10.1021/ol026752a

58. Kale, T. A.; Hsieh, S.-A. J.; Rose, M. W.; Distefano, M. D., Use of synthetic isoprenoid analogues for understanding protein prenyltransferase mechanism and structure. *Current Topics in Medicinal Chemistry* 2003, 3, (10), 1043-1074. DOI: 10.2174/1568026033452087

57. Kale, T. A.; Raab, C.; Yu, N.; Aquino, E.; Dean, D. C.; Distefano, M. D., Synthesis of high specific activity ^{35}S-labelled N-methanesulfonyl farnesylcysteine and a photoactive analog. *Journal of Labelled Compounds & Radiopharmaceuticals* 2003, 46, (1), 29-54. DOI: 10.1002/jlcr.638

56. Turek-Etienne, T. C.; Strickland, C. L.; Distefano, M. D., Biochemical and Structural Studies with Prenyl Diphosphate Analogues Provide Insights into Isoprenoid Recognition by Protein Farnesyl Transferase. *Biochemistry* 2003, 42, (13), 3716-3724. DOI: 10.1021/bi0266838

55. Distefano, M. D.; Zhong, A.; Cochran, A. G., Quantifying â-Sheet stability by phage display. *Journal of Molecular Biology* 2002, 322, (1), 179-188. DOI:10.1016/S0022-2836(02)00738-6

54. Häring, D.; Lees, M. R.; Banaszak, L. J.; Distefano, M. D., Exploring routes to stabilize a cationic pyridoxamine in an artificial transaminase: site-directed mutagenesis versus synthetic cofactors. *Protein Engineering* 2002, 15, (7), 603-10.

53. Kale, T. A.; Turek, T. C.; Chang, V.; Gautam, N.; Distefano, M. D., Preparation and application of G protein g subunit-derived peptides incorporating a photoactive isoprenoid. *Methods in Enzymology* 2002, 344, (G Protein Pathways, Part B), 245-258.

52. Clausen, V. A.; Edelstein, R. L.; Distefano, M. D., Stereochemical Analysis of the Reaction Catalyzed by Human Protein Geranylgeranyl Transferase. *Biochemistry* 2001, 40, (13), 3920- 3930. DOI: 10.1021/bi002011a

51. Häring, D.; Distefano, M. D., Enzymes by Design: Chemogenetic Assembly of Transamination Active Sites Containing Lysine Residues for Covalent Catalysis. *Bioconjugate Chemistry* 2001, 12, (3), 385-390. DOI: 10.1021/bc000117c

50. Häring, D.; Distefano, M. D., Specific Host-Guest Interactions in a Protein-Based Artificial Transaminase. *Bioorganic and Medicinal Chemistry* 2001, 9, 2461-2466.

49. Häring, D.; Distefano, M. D., Converting a Fatty Acid Binding Protein to a Artificial Transaminase: Novel Catalysts by Chemical and Genetic Modification of a Protein Cavity. *Journal of Molecular Catalysis. B, Enzymatic* 2001, 11, 967-970.
48. Kale, T. A.; Raab, C.; Yu, N.; Dean, D. C.; Distefano, M. D., A Photoactivatable Prenylated Cysteine Designed to Study Isoprenoid Recognition. *Journal of the American Chemical Society* 2001, 123, (19), 4373-4381. DOI: 10.1021/ja0012016
47. Qi, D.; Tann, C.-M.; Häring, D.; Distefano, M. D., Generation of new enzymes via covalent modification of existing proteins. *Chemical Reviews* 2001, 101, (10), 3081-3111. DOI: 10.1021/cr000059o.
46. Tann, C.-M.; Qi, D.; Distefano, M. D., Enzyme design by chemical modification of protein scaffolds. *Current Opinion in Chemical Biology* 2001, 5, (6), 696-704.
45. Turek, T. C.; Gaon, I.; Distefano, M. D.; Strickland, C. L., Synthesis of Farnesyl Diphosphate Analogues Containing Ether-Linked Photoactive Benzophenones and Their Application in Studies of Protein Prenyltransferases. *Journal of Organic Chemistry* 2001, 66, (10), 3253-3264. DOI: 10.1021/jo991130x.
44. Distefano, M. D.; Kuang, H.; Qi, D.; Häring, D.; Ory, J.; Banaszak, L. J., Semisynthetic approaches for the design of proteins with catalytic activity using fatty acid binding protein as a scaffold. *Peptides for the New Millennium, Proceedings of the American Peptide Symposium, 16th, Minneapolis, MN, United States, June 26-July 1, 1999* 2000, 500-502.
43. Kale, T. A.; Distefano, M. D., Photoaffinity-labeled probes for the study of isoprenoid recognition sites. *Peptides for the New Millennium, Proceedings of the American Peptide Symposium, 16th, Minneapolis, MN, United States, June 26-July 1, 1999* 2000, 109-110.
42. Kuang, H.; Häring, D.; Qi, D.; Mazhary, A.; Distefano, M. D., Synthesis of a cationic pyridoxamine conjugation reagent and application to the mechanistic analysis of an artificial transaminase. *Bioorganic & Medicinal Chemistry Letters* 2000, 10, 2091-2095.
41. Davies, R. R.; Kuang, H.; Qi, D.; Mazhary, A.; Mayaan, E.; Distefano, M. D., Artificial metalloenzymes based on protein cavities: exploring the effect of altering the metal ligand attachment position by site directed mutagenesis. *Bioorganic & Medicinal Chemistry Letters* 1999, 9, (1), 79-84.
40. Distefano, M. D.; Kuang, H.; Qi, D.; Mazhary, A., The design of protein-based catalysts using semisynthetic methods. *Current Opinion in Structural Biology* 1998, 8, (4), 459-465.
39. Edelstein, R. L.; Weller, V. A.; Distefano, M. D.; Tung, J. S., Stereochemical Analysis of the Reaction Catalyzed by Yeast Protein Farnesyltransferase. *Journal of Organic Chemistry* 1998, 63, (16), 5298-5299. DOI: 10.1021/jo980304s
38. Kuang, H.; Distefano, M. D., Catalytic Enantioselective Reductive Amination in a Host-Guest System Based on a Protein Cavity. *Journal of the American Chemical Society* 1998, 120, (5), 1072-1073. DOI: 10.1021/ja972771h
37. Ory, J. J.; Mazhary, A.; Kuang, H.; Davies, R. R.; Distefano, M. D.; Banaszak, L. J., Structural characterization of two synthetic catalysts based on adipocyte lipid-binding protein. *Protein Engineering* 1998, 11, (4), 253-261.

36. Qi, D.; Kuang, H.; Distefano, M. D., Effects of metal ions on the rates and enantioselectivities of reactions catalyzed by a series of semisynthetic transaminases created by site directed mutagenesis. *Bioorganic & Medicinal Chemistry Letters* 1998, 8, (7), 875-880.
35. Tabakovic, I.; Gaon, I.; Distefano, M. D., Anodic oxidation of substituted [4-hydroxy-3-coumarinyl]-phenylthiomethanes in acetonitrile. Evidence for a cationic intermediate in carbon-sulfur bond fragmentation. *Electrochimica Acta* 1998, 43, (12-13), 1773-1778.
34. Turek, T. C.; Edelstein, R. L.; Gaon, I.; Weller, V. A.; Distefano, M. D., [^{32}P]-labeled analogs of farnesyl and geranylgeranyl pyrophosphate: Synthesis and application in photoaffinity labeling experiments with protein prenyltransferases. In *Synthesis and Applications of Isotopically Labelled Compounds 1997, Proceedings of the International Symposium, 6th, Philadelphia, PA, United States, Sept. 14-18, 1997*, 1998; p 67-70.
33. Weller, V. A.; Distefano, M. D., Measurement of the á-Secondary Kinetic Isotope Effect for a Prenyltransferase by MALDI Mass Spectrometry. *Journal of the American Chemical Society* 1998, 120, (31), 7975-7976. DOI: 10.1021/ja980353m
32. Davies, R. R.; Distefano, M. D., A Semisynthetic Metalloenzyme Based on a Protein Cavity That Catalyzes the Enantioselective Hydrolysis of Ester and Amide Substrates. *Journal of the American Chemical Society* 1997, 119, (48), 11643-11652. DOI: 10.1021/ja970820k
31. Edelstein, R. L.; Distefano, M. D., Photoaffinity Labeling of Yeast Farnesyl Protein Transferase and Enzymatic Synthesis of a Ras Protein Incorporating a Photoactive Isoprenoid. *Biochemical and Biophysical Research Communications* 1997, 235, 377-382.
30. Herbert, C. B.; McLernon, T. L.; Hypolite, C. L.; Adams, D. N.; Pikus, L.; Huang, C. C.; Fields, G. B.; Letourneau, P. C.; Distefano, M. D.; Hu, W.-S., Micropatterning gradients and controlling surface densities of photoactivatable biomolecules on self-assembled monolayers of oligo(ethylene glycol) alkanethiolates. *Chemistry & Biology* 1997, 4, (10), 731-737.
29. Hypolite, C. L.; McLernon, T. L.; Adams, D. N.; Chapman, K. E.; Herbert, C. B.; Huang, C. C.; Distefano, M. D.; Hu, W.-S., Formation of Microscale Gradients of Protein Using Heterobifunctional Photolinkers. *Bioconjugate Chemistry* 1997, 8, (5), 658-663. DOI: 10.1021/bc9701252
28. Kuang, H.; Davies, R. R.; Distefano, M. D., Modulation of the rate, enantioselectivity, and substrate specificity of semisynthetic transaminases based on lipid binding proteins using site directed mutagenesis. *Bioorganic & Medicinal Chemistry Letters* 1997, 7, (15), 2055-2060.
27. Turek, T. C.; Gaon, I.; Distefano, M. D., Synthesis and rapid purification of ^{32}P-labeled photoactive analogs of farnesyl pyrophosphate. *Journal of Labelled Compounds & Radiopharmaceuticals* 1997, 39, (2), 139-146.
26. Turek, T. C.; Gaon, I.; Gamache, D.; Distefano, M. D., Synthesis and evaluation of benzophenone-based photoaffinity labeling analogs of prenyl pyrophosphates containing stable amide linkages. *Bioorganic & Medicinal Chemistry Letters* 1997, 7, (16), 2125-2130.

25. Gaon, I.; Turek, T. C.; Distefano, M. D., Farnesyl and geranylgeranyl pyrophosphate analogs incorporating benzoylbenzyl ethers: synthesis and inhibition of yeast protein farnesyltransferase. *Tetrahedron Letters* 1996, 37, (49), 8833-8836.
24. Gaon, I.; Turek, T. C.; Weller, V. A.; Edelstein, R. L.; Singh, S. K.; Distefano, M. D., Photoactive Analogs of Farnesyl Pyrophosphate Containing Benzoylbenzoate Esters: Synthesis and Application to Photoaffinity Labeling of Yeast Protein Farnesyltransferase. *Journal of Organic Chemistry* 1996, 61, (22), 7738-7745. DOI: 10.1021/jo9602736
23. Kuang, H.; Brown, M. L.; Davies, R. R.; Young, E. C.; Distefano, M. D., Enantioselective Reductive Amination of á-Keto Acids to á-Amino Acids by a Pyridoxamine Cofactor in a Protein Cavity. *Journal of the American Chemical Society* 1996, 118, (44), 10702-10706. DOI: 10.1021/ja954271z
22. Turek, T. C.; Gaon, I.; Distefano, M. D., Analogs of farnesyl pyrophosphate incorporating internal benzoylbenzoate esters: synthesis, inhibition kinetics and photoinactivation of yeast protein farnesyltransferase. *Tetrahedron Letters* 1996, 37, (28), 4845-4848.
21. Colocci, N.; Distefano, M. D.; Dervan, P. B., Cooperative oligonucleotide-directed triple helix formation at adjacent DNA sites. *Journal of the American Chemical Society* 1993, 115, (11), 4468-73. DOI: 10.1021/ja00064a008
20. Distefano, M. D.; Dervan, P. B., Energetics of cooperative binding of oligonucleotides with discrete dimerization domains to DNA by triple helix formation. *Proceedings of the National Academy of Sciences of the United States of America* 1993, 90, (4), 1179-83.
19. Distefano, M. D.; Dervan, P. B., Ligand-promoted dimerization of oligonucleotides binding cooperatively to DNA. *Journal of the American Chemical Society* 1992, 114, (27), 11006-7. DOI: 10.1021/ja00053a073
18. Distefano, M. D.; Shin, J. A.; Dervan, P. B., Cooperative binding of oligonucleotides to DNA by triple helix formation: dimerization via Watson-Crick hydrogen bonds. *Journal of the American Chemical Society* 1991, 113, (15), 5901-2. DOI: 10.1021/ja00015a076
17. Schiering, N.; Kabsch, W.; Moore, M. J.; Distefano, M. D.; Walsh, C. T.; Pai, E. F., Structure of the Detoxification Catalyst Mercuric Ion Reductase from Bacillus sp. strain RC607. *Nature* 1991, 352, 168-172.
16. Schiering, N.; Kabsch, W.; Moore, M. J.; Distefano, M. D.; Walsh, C. T.; Pai, E. F., The Three Dimensional Structure of Mercuric Ion Reductase from Bacillus Strain RC607. In *Flavins and Flavoproteins 1990*, ed.; Curti, B.; Ronchi, S.; Zanetti, G., Walter de Gruyter & Co.: Berlin, 1991; p 615-625.
15. Distefano, M. D.; Moore, M. J.; Walsh, C. T., Active site of mercuric reductase resides at the subunit interface and requires Cys135 and Cys140 from one subunit and Cys558 and Cys559 from the adjacent subunit: evidence from in vivo and in vitro heterodimer formation. *Biochemistry* 1990, 29, (11), 2703-13. DOI: 10.1021/bi00463a013
14. Miller, S. M.; Massey, V.; Ballou, D.; Williams, C. H., Jr.; Distefano, M. D.; Moore, M. J.; Walsh, C. T., Use of a site-directed triple mutant to trap intermediates: demonstration that the flavin C(4a)-thiol adduct and reduced flavin are kinetically competent intermediates in mercuric ion reductase. *Biochemistry* 1990, 29, (11), 2831-41. DOI: 10.1021/bi00463a028

13. Moore, M. J.; Distefano, M. D.; Zydowsky, L. D.; Cummings, R. T.; Walsh, C. T.,Organomercurial lyase and mercuric ion reductase: nature's mercury detoxification catalysts. *Accounts of Chemical Research* 1990, 23, (9), 301-8. DOI: 10.1021/ar00177a006

12. Raybuck, S. A.; Distefano, M. D.; Teo, B. K.; Orme-Johnson, W.; Walsh, C. T., An EXAFS investigation of Hg(II) binding to mercuric reductase: comparative analysis of the wild-type enzyme and a mutant enzyme generated by site-directed mutagenesis. *Journal of the American Chemical Society* 1990, 112, (5), 1983-9. DOI: 10.1021/ja00161a051

11. Distefano, M. D.; Au, K. G.; Walsh, C. T., Mutagenesis of the redox-active disulfide in mercuric ion reductase: catalysis by mutant enzymes restricted to flavin redox chemistry. *Biochemistry* 1989, 28, (3), 1168-83. DOI: 10.1021/bi00429a035

10. Miller, S. M.; Moore, M. J.; Massey, V.; Williams, C. H., Jr.; Distefano, M. D.; Ballou, D. P.; Walsh, C. T., Evidence for the participation of Cys558 and Cys559 at the active site of mercuric reductase. *Biochemistry* 1989, 28, (3), 1194-205. DOI: 10.1021/bi00429a037

9. Moore, M. J.; Distefano, M. D.; Walsh, C. T.; Schiering, N.; Pai, E. F., Purification, crystallization, and preliminary x-ray diffraction studies of the flavoenzyme mercuric ion reductase from *Bacillus sp. strain RC607. Journal of Biological Chemistry* 1989, 264, (24), 14386-8.

8. Walsh, C.; Distefano, M. D.; Moore, M., Mutagenesis of paired cysteine residues in the disulphide-containing flavoprotein mercuric ion reductase from mercury-resistant bacteria. *Biochemical Society Transactions* 1988, 16, (2), 90-1.

7. Walsh, C. T.; Distefano, M. D.; Moore, M. J.; Shewchuk, L. M.; Verdine, G. L., Molecular basis of bacterial resistance to organomercurial and inorganic mercuric salts. *FASEB Journal* 1988, 2, (2), 124-30.

6. Massey, V.; Miller, S.; Ballou, D. P.; Williams, C. H.; Moore, M. J.; Distefano, M. D.; Walsh, C. T., Studies of the Active Site of Mercuric Reductase Employing Site Directed Mutants and the Thiol-Reactive Flavin, 6-SCN-FAD. In *Flavins and Flavoproteins 1987*, ed.; Edmondson, D. E.; McCormick, D. B., Walter de Gruyter & Co.: Berlin, 1987; p 41-44.

5. Moore, M. J.; Distefano, M. D.; Walsh, C. T.; Miller, S.; Massey, V.; Williams, C. H.; Ballou, D. P., The Penultimate Cysteines in Mercuric Reductase Aid in the Reduction of Mercury. In *Flavins and Flavoproteins 1987*, ed.; Edmondson, D. E.; McCormick, D. B., Walter de Gruyter & Co.: Berlin, 1987; p 37-40.

4. Neidhard, D. J.; Distefano, M. D.; Tanizawa, K.; Soda, K.; Walsh, C. T.; Petsko, G. A., X-ray crystallographic studies of the alanine-specific racemase from *Bacillus stearothermophilus*. Overproduction, crystallization, and preliminary characterization. *Journal of Biological Chemistry* 1987, 262, (32), 15323-6.

3. Walsh, C.; Distefano, M. D.; Moore, M., Catalytic effects of mutagenesis of paired cysteine residues in the bacterial enzyme mercuric ion reductase. *UCLA Symposia on Molecular and Cellular Biology, New Series* 1987, 69, (Protein Struct., Folding, Des. 2), p 283-92.

2. Walsh, C. T.; Moore, M. J.; Distefano, M. D., Conserved Cysteine Pairs of Mercuric Ion Reductase: An Investigation of Function Via Site-Directed Mutagenesis. In *Flavins and Flavoproteins 1987*, ed.; Edmondson, D. E.; McCormick, D. B.,Walter de Gruyter & Co.: Berlin, 1987; p 13-28.
1. Kanazawa, T.; Distefano, M. D.; Bassham, J. A., Ammonia regulation of intermediary metabolism in photosynthesizing and respiring *Chlorella pyrenoidosa*: comparative effects of methylamine. *Plant and Cell Physiology* 1983, 24, (6), 979-86.

In: Biochemistry Researcher Biographical Sketches ...
Editors: J. T. Walker and C. M. Jones
ISBN: 978-1-62081-308-9

Chapter 8

Azza Ali Kamel El-Sheikh

Affiliation:

Pharmacology Dep., Faculty of Medicine, Minia University, Egypt

Address:

Pharmacology Dep., Faculty of Medicine, Minia University, 61511 Minia, Egypt.

Date of Birth:

23/11/1969

Education:

- PhD of Pharmacology; Radboud University Nijmegen, the Netherlands (2008).
- Master of Pharmacology; Minia University, Egypt (1998).
- Bachelor of Medicine and Surgery; Minia University, Egypt (1993).

Research and Professional Experience:

A. Technical skills:

1. General skills:

- Excellent computer skills (ICDL in 2004 from Cairo University, Egypt).
- International computer-based TOEFL Exam
- Computer typing skills: 40 words per minute (Arabic and English).

2. Laboratory Skills:

- Experience of radiolabelled compound handling acquired during Doctoral degree in Radboud University Nijmegen, the Netherlands (Holland).

- Experience of some laboratory techniques during studying for the Master and Doctoral degrees as cell culturing, virus-induced transduction for protein over-expression, cell membrane vesicles isolation, transport/inhibition studies using radiolabelled compounds, drug monitoring using high performance liquid chromatography, western blotting, immunohistochemical staining of tissues and cells, spectrophotometric and colorimetric tests.
- Experience of animal handling during Master degree, performed on Albino rabbits.

B. Courses attended:

1- Faculty and Leadership Development Project (FLDP) training courses, Minia University, Egypt (2009), included courses in the area of quality management systems as:

- Quality academic standards in teaching process
- Strategic planning
- Communication skills in different types of education
- Educational professional ethics
- Financial and legal rules of the university
- Effective educational presentation

2- NCMLS PhD Research training program (2004-2008), and receiving the NCMLS Diploma, including:

- PhD student orientation course
- NCMLS Seminars, forums, and PhD workshops
- NCMLS PhD student yearly 2-day conferences
- Participating in weekly department meetings
- Participating in monthly colloquiums

3- The Seventh Dutch Nephrology Winter School course, organized by the Dutch Kidney Foundation (2007) held in Zandvoort, the Netherlands (Holland).

4- Radiation Hygiene course and examination (2007), held in Wageningen, the Netherlands (Holland).

5- Immunology Master course and examination: Infection, immunity and tissue repair theme (2005) in Radboud University Nijmegen, the Netherlands (Holland).

C. Present Duties:

- Contribution in teaching Pharmacology to undergraduate students from Faculties of Medicine, Dentist and Nursing in Minia University, Egypt.
- Contributing in Pharmacology Master thesis co-supervision and teaching postgraduate Pharmacology course.
- Coordinator of environmental affairs in Department of Pharmacology, which includes duties in organizing work spaces, coordination of the department seminars, contributing to the CIQAP project and coordinating department information on the university website and IT affairs.
- Pharmacology course coordinator of Nursing Institute.
- Member of the Egyptian Society of Pharmacology and Experimental Therapeutics

Professional Appointments:

- Assistant Prof. in Pharmacology, Faculty of Medicine, Minia University, Egypt (2009-present).
- Assistant lecturer of Pharmacology, Faculty of Medicine, Minia University, Egypt (2008-2009).
- PhD researcher in Pharmacology, Radboud University Nijemgen, the Netherlands (2004-2008).
- Assistant lecturer of Pharmacology, Faculty of Medicine, Minia University, Egypt (1999-2004).
- Demonstrator of Pharmacology, Faculty of Medicine Minia University, Egypt (1995-1999).
- General Practitioner in Mataria Educational Hospital, Cairo, Egypt (1994-1995).

Honors:

1- Best oral presentation certificate during the "Drug-related research PhD competition" from Radboud University Nijmegen, and elected to participate in the Dutch National PhD competition During Figon Conference 2006, the Netherlands (Holland). This participation was quoted in the "Conceptuur" magazine, in December 2006 issue, volume 49, page 10 (article in Dutch), URL: http://www.figon.nl/pdf/conceptuur-49.pdf

2- Best discussion prize during the "Nijmegen Center for Molecular Life Sciences PhD 2-day conference" held on 27th-28th April 2006 at the National Sports Center, Papendal Hotel, Arnhem, the Netherlands (Holland).

3- Quoted in the Women International day in Radboud University Nijmegen Vox magazine, 10th edition, vol. 13 (article in Dutch). URL: http://www.ru.nl/vox/archief/zoek_op_nummer/jaargang-7/nummer_13/artikelen/alles_voor_de/

Scholarship sponsoring a 4-years PhD research in the Netherlands, granted by the Egyptian Ministry of higher education in 2003.

Publications Last 3 Years:

A- Books:

1- Mohamed A. Morsy, Azza A.K. El-Sheikh. Prevention of Gastric Ulcers. In: Peptic Ulcer Disease. Chai J (Ed.), InTech, Croatia, 2011. ISBN 978-953-307-976-9 http://www.intechopen.com/articles/show/title/prevention-of-gastric-ulcers

2- Azza A.K. El-Sheikh. Pharmacology of the central nervous system. In: Practical and Clinical Pharmacology Notes for undergraduate pharmacology students, Department of Pharmacology, Faculty of Medicine, Minia University, Egypt, 2009.

3- Azza A.K. El-Sheikh. Immuno-pharmacology. In: Practical and Clinical Pharmacology Notes for undergraduate pharmacology students, Department of Pharmacology, Faculty of Medicine, Minia University, Egypt, 2009.

4- Azza A.K. El-Sheikh. Autacoids. In: Practical and Clinical Pharmacology Notes for undergraduate pharmacology students, Department of Pharmacology, Faculty of Medicine, Minia University, Egypt, 2009.

5- Azza A.K. El-Sheikh. PhD doctoral thesis Pharmacology book entitled: "Renal Transport and Drug Interactions of Immunosuppressants". (ISBN: 978-90-9023458-8). PhD research was conducted in Radboud University Nijmegen in the Netherlands (Holland). The URL link for the thesis is http://webdoc.ubn.ru.nl/mono/e/el_sheikh_a/renatrand.pdf , 2008.

6- Azza A.K. El-Sheikh Master thesis Pharmacology book entitled: "Possible Antioxidant Effect of SH-Containing Angiotensin Converting Enzyme Inhibitors in Ameliorating Atherogenesis". Master research was conducted in Minia University, Egypt, 1998.

B- Journal articles:

1. Azza A.K. El-Sheikh, Mohamed A. Morsy, Marwa M. Mahmoud, Rehab A. Rifaai and Aly M. Abdelrahman. Differential Dose-dependent Mechanisms of Coenzyme Q10 in Doxorubicin-induced Nephrotoxicity in Rats. Food and Chem. Tox., 2011 (submitted)
2. Azza A.K. El-Sheikh, Jeroen J.M.W. van den Heuvel, Jan B. Koenderink, and Frans G.M. Russel. Characterization of human URAT1 in HEK293 Cells and membrane vesicles. *Pharm. Res.*, 2011 (prepared for submission).
3. Azza A.K. El-Sheikh, Jeroen J.M.W. van den Heuvel, Jan B. Koenderink, and Frans G.M. Russel. Interactions of Immunosuppressants with Human Organic Anion Transporter 1 and Multidrug Resistance Proteins 2 and 4. *Br. J. Pharm.* 2011 (prepared for submission).
4. Azza A.K. El-Sheikh, Jeroen J.M.W. van den Heuvel, Jan B. Koenderink, and Frans G.M. Russel. Characterization of Human URAT1 Exchanger in HEK293 Cells and Membrane Vesicles. *Eur. J. Pharm.*, 2011 (prepared for submission).
5. Azza A.K. El-Sheikh, Alfons C. Wouterse, Petra H.H. van den Broek, Rosalinde Masereeuw and Frans G.M. Russel: Cyclosporine A inhibits renal mycophenolic acid glucuronidation and multidrug resistance protein 2-mediated excretion. *Kid. Intern.*, 2011 (prepared for submission).
6. Azza A.K. El-Sheikh, Jeroen J.M.W. van den Heuvel, Jan B. Koenderink, and Frans G.M. Russel: Effect of hypouricemic and hyperuricemic drugs on renal urate efflux transporter multidrug resistance protein (MRP) 4. *Br. J. Pharmacol.*, 155: 1066-1075, 2008.
7. Azza A.K. El-Sheikh, Jeroen J.M.W. van den Heuvel, Elmar Krieger, Frans G.M. Russel, and Jan B. Koenderink: Functional role of Arginine 375 in transmembrane helix 6 of multidrug resistance protein 4 (MRP4/ABCC4). *Mol. Pharmacol.*, 74: 964-971, 2008.
8. Azza A.K. El-Sheikh, Rosalinde Masereeuw and Frans G.M. Russel: Mechanisms of renal anionic drug transport. *Eur. J. Pharmacol.*, 585: 245-255, 2008.
9. Azza A.K. El-Sheikh, Jeroen J.M.W. van den Heuvel, Jan B. Koenderink, and Frans G.M. Russel: Interaction of nonsteroidal anti-inflammatory drugs with multidrug resistance protein (MRP) 2/ABCC2- and MRP4/ABCC4-mediated methotrexate transport. *J. Pharmacol. Exp. Ther.*, 320: 229-35, 2007.

10. Nijmegen Center for Molecular Life Sciences Annual Report entitled "U*nderstanding Renal Drug-Drug Interactions*", page 40, 2007.
11. www.ncmls.eu/ncmls/docs/Annualreports/NCMLS%202007.pdf

C- Abstracts:

1. Azza A.K. El-Sheikh, Alfons C. Wouterse, Petra H.H. van den Broek, Rosalinde Masereeuw and Frans G.M. Russel: Cyclosporin A but not tacrolimus inhibits renal mycophenolic acid glucuronidation and multidrug resistance protein (MRP) 2-mediated excretion. *AAPS Annual Meeting and Exposition* (abstract), 2009.
2. Azza A.K. El-Sheikh, Jeroen J.M.W. van den Heuvel, Jan B. Koenderink, and Frans G.M. Russel. Interaction of immunosuppressants with methotrexate transport mediated by human multidrug resistance transporters MRP2 and MRP4. Naunyn- Schmiedebergs *Archives of Pharmacology* (abstract), 379: 202, 2009.
3. Azza A.K. El-Sheikh, Jeroen J.M.W. van den Heuvel, Jan B. Koenderink, and Frans G.M. Russel. Interaction of uricosurics and diuretics with the renal urate efflux transporter MRP4/ABCC4. Naunyn-Schmiedebergs *Archives of Pharmacology* (abstract) 375: 148, 2007.
4. Azza A.K. El-Sheikh, Jeroen J.M.W. van den Heuvel, Jan B. Koenderink, and Frans G.M. Russel. Interaction of NSAIDs with MRP2/ABCC2-and MRP4/ABCC4-mediated methotrexate transport. *Naunyn-Schmiedebergs Archives of Pharmacology* (abstract) 373: 92, 2006.
5. Mohamed M.A. Khalifa, Magdy K.A. Hassan, Osama M. Ashour, Mohamed A. Ibrahim, Azza A.K. El-Sheikh and Abdel-Baset T.A. Abdel-Haleem. Lipid peroxidation and Hypercholesterolemic Atherosclerosis: Effect of some calcium channel blockers and some Angiotensin converting enzyme inhibitors. *The 42nd Annual Scientific Conference of Al-Azhar University, supervised by The Egyptian Society of Pharmacology and Experimental Therapeutics,* (abstract), 1998**.**

D- Bachelor and Master co-supervision:

1. Master (M.Sc.) Thesis of Mrs. Marwa M. Refaie, entitled: The protective effect of antioxidants on doxorubicin nephrotoxicity, at the Department of Pharmacology, Faculty of Medicine, El-Minia University, 2010.
2. Bachelor internship of Ms. Lianne Bos, entitled: Interactions of Immunosuppressant Drugs with Human Renal Basolateral Organic Anion transporters, at the Deparmtment of Pharmacology and Toxicology, Radboud University Nijmegen, the Netherlands (Holland), 2008.
3. Bachelor internship of Ms. Maarke M.J.E. Roelofs, entitled: Drug interactions at Human Renal Proximal Tubules Apical Transporters MRP4 and URAT1, at the Deparmtment of Pharmacology and Toxicology, Radboud University Nijmegen, the Netherlands (Holland), 2007.
4. Master internship of Iris M.J. Dros, entitled: Characterization of Mycophenolic Acid and its Metabolite as a Substrate of Human MRP2 Transporter, at the Deparmtment of Pharmacology and Toxicology, Radboud University Nijmegen, the Netherlands (Holland), 2007.

5. Paper writing supervision of I.H.J. Ploemen and K. van den Dries, entitled: Thiazides, uricosurics and xanthine oxidase inhibitors: interactions with urate transport mediated by the multidrug resistance protein MRP4, at the Deparmtment of Pharmacology and Toxicology, Radboud University Nijmegen, the Netherlands (Holland), 2006.
6. Paper writing supervision of H.S. Hendriks and W.M. Rensen van den Dries, entitled: Influence of NSAIDs on Methotrexate Transport by Multidrug Resistance Proteins MRP2 and MRP4, at the Deparmtment of Pharmacology and Toxicology, Radboud University Nijmegen, the Netherlands (Holland), 2005.

In: Biochemistry Researcher Biographical Sketches …
Editors: J. T. Walker and C. M. Jones
ISBN: 978-1-62081-308-9

Chapter 9

Boris Estrine

Affiliation:

Head of Green Chemistry Department (ARD).

Address:

b.estrine@a-r-d.fr

Date of Birth:

3rd of January 1976

Education:

PhD in organic chemistry in 2002

Research and Professional Experience:

Head of Green Chemistry Department (ARD). Current interests: Development of products and processes from biomass resources following green chemistry guidelines. Activities lie in the field of surfactant, solvent and polymers/materials. Specific knowledge in metal catalyzed reactions, carbohydrate chemistry and product formulation for detergents, cosmetics and miscellaneous industrial sectors

Publications Last 3 Years:

Novel use of biomass derived alkyl-xylosides in wetting agent for paper impregnation suitable for the wood-based industry, Papadopoulou, E., Hatjiissaak, A., Estrine, B., Marinkovic, S. *European Journal of Wood and Wood Products* 2011, 1-7 In press.

Direct conversion of xylan into alkyl pentosides, Bouxin, F., Marinkovic, S., Bras, J.L., Estrine, B. *Carbohydrate Research* 2010, *345*, 2469-2473.

Direct conversion of wheat bran hemicelluloses into n-decyl-pentosides, Marinkovic, S., Estrine, B. *Green Chem.* 2010, *12*, 1929-1932.

Development of agriculture left-overs: Fine organic chemicals from wheat hemicellulose-derived pentoses, Martel, F., Estrine, B., Plantier-Royon, R., Hoffmann, N., Portella, C. *Topics in Current Chemistry*, 2010, *294*, 79-115.

Low catalyst loading for the production of carboxylic acids from polysaccharides and hydrogen peroxide, C. Aouf, D. Harakat, J. Le Bras, J. Muzart, B. Estrine, S. Marinkovic, C. Ernenwein *ChemSusChem* 2010, *3*, 1200-1203.

Recycling in telomerization of butadiene with D-xylose: Pd(TPPTS) n-KF/Al2O3 as an active catalyst, Estrine, B., Bouquillon, S., Hénin, F., Muzart, J. *Applied Organometallic Chemistry*, 2007, *21*, 945-946.

Emulsifying composition comprising alkyl polyglycosides and derivatives of succinic acid. Renault, B., Estrine, B., Portella, C., Ernenwein, C., Massicot, F., Marinkovic, S. patent GB2468575. 2010 .

Novel alkyl glycoside-based compositions, preparation method thereof and use of the same as surfactants. Estrine, B., Marinkovic, M., Ernenwein, C. patents WO08135646. 2008 . and EP2139594. 2010.

Novel emulsifiers and the use thereof for preparing vaporizable emulsions. Ernenwein, C., Estrine, B. patent EP1750834. 2009.

In: Biochemistry Researcher Biographical Sketches ...
Editors: J. T. Walker and C. M. Jones
ISBN: 978-1-62081-308-9

Chapter 10

Túlio César Ferreira

Affiliation:

Universidade de Brasília – Departamento de Biologia Celular

Address:

Universidade de Brasília
Departamento de Biologia Celular
Laboratório de Biologia Molecular
Campus Darcy Ribeiro – Asa Norte
Brasília-DF - Brasil
CEP 70910-900

Date of Birth:

28th January, 1976

Research and Professional Experience:

Molecular Biology and Biochemistry focused on (1) oxidative stress and antioxidant mechanisms in yeast cells, (2) studies of antioxidant properties of plants and (3) studies on role of hypoxia-inducible factor-1 (HIF-1). Teaching in Biochemistry.

Publications Last 3 Years: 2

In: Biochemistry Researcher Biographical Sketches ... ISBN: 978-1-62081-308-9
Editors: J. T. Walker and C. M. Jones

Chapter 11

Rose Gasnier

Affiliation:

Hospital de Clinicas de Porto Alegre, Post-Graduate Program in Medicine, Universidade Federal do Rio Grande do Sul

Address:

Rua Ramiro Barcelos 1691/11, 90035-006 Porto Alegre, RS, Brazil.
Tel +55 51 33325173

Date of Birth: 31/03/1979

Research and Professional Experience:

Obstetrics and Gynecology
Master Degree — Calciuria in Induced-Pregnancy Hypertension

Professional Appointments:

Obstetrician and Gynecologist

Publications Last 3 Years:

Calciuria in pregnancy-induced hypertension. Master Degree Dissertation, 2010. UFRGS Digital Repository, Universidade Federal do Rio Grande do Sul.

Gasnier R, Valerio EG, Vettorazzi J, Martins-Costa SH, Barros EG, Ramos JGL. Calcium-to-creatinine ratio in pregnancy-induced hypertension. *Pregnancy Hypertension: An International Journal of Women's Cardiovascular Health* 2012(2): 59-64.

Gasnier R, Valerio EG, Vettorazzi J, Martins-Costa SH, Barros EG, Ramos JGL. Calciuria and preeclampsia: a case-control study. *J Obst Gynecol Res* 2012 — accepted

Muller ALL, Barrios PM, Kliemann LM, Valerio EG, Gasnier R, Magalhaes JAA. The tei index to assess fetal cardiac performance in fetus at risk for fetal inflammatory response syndrome. *Ultrasound in Obstetrics and Gynecology* 2010(9999A).

Gasnier R, Valerio EG, Vettorazzi J, Martins-Costa SH, Barros EG, Ramos JGL. Calciuria in pregnancy-induced hypertension and perinatal outcomes. In: XVII ISSHP World Congress, Melbourne, Australia, 3-6 October 2010. *Pregnancy Hypertension: An International Journal of Women's Cardiovascular Health 2010(1):* 544. Abstracts of the XVII ISSHP World Congress.

Gasnier R, Valerio EG, Vettorazzi J, Martins-Costa SH, Barros EG, Ramos JGL. Calcium-to-creatinine ratio in preeclapsia. In: XVII ISSHP World Congress, Melbourne, Australia, 3-6 October 2010. *Pregnancy Hypertension: An International Journal of Women's Cardiovascular Health 2010(1): 544*. Abstracts of the XVII ISSHP World Congress.

Gasnier R, Valerio EG, Vettorazzi J, Martins-Costa SH, Barros EG, Ramos JGL. Natriuria associated to calciuria in pregnancy-induced hypertension. In: XVII ISSHP World Congress, Melbourne, Australia, 3-6 October 2010. *Pregnancy Hypertension: An International Journal of Women's Cardiovascular Health 2010(1): 544*. Abstracts of the XVII ISSHP World Congress.

In: Biochemistry Researcher Biographical Sketches ... ISBN: 978-1-62081-308-9
Editors: J. T. Walker and C. M. Jones

Chapter 12

Suresh Kumar Kailasa

Affiliation:

S. V. National Institute of Technology, Surat – 395007, India

Address:

Department of Applied Chemistry, S. V. National Institute of Technology, Surat – 395007, India

Date of Birth:

15.06.1977

Education:

Ph.D

Research and Professional Experience:

Nanochemistry, Mass Spectrometry and Proteomics; Organic and Analytical Chemistry

Professional Appointments:

Assistant Professor, SVNIT, Surat, India

Honors:

Young Scientist, Department of Science and Technology, INDIA

Publications Last 3 Years:

1. K. Suresh Kumar, Hui-Fen Wu, Semiconductor cadmium sulphide nanoparticles as matrices for peptides and as co-matrices for the analysis of large proteins in MALDI reflectron and linear time-of-flight mass spectrometry, *Rapid Communications in Mass Spectrometry,* 25 (2), 271–280, 2011.
2. Hui-Fen Wu, K. Suresh Kumar, Chi-Hsien Lin, Single drop microextraction coupled with MALDI mass spectrometry for the rapid and direct analysis of hydrophobic peptides from biological samples at high salt condition, *Rapid Communications in Mass Spectrometry,* 25 (2), 307–315, 2011.
3. K. Suresh Kumar, Hui-Fen Wu, Surface modified silver selinide nanoparticles as extracting probes to improve peptide/protein detection via nanoparticles-based liquid phase microextraction coupled with MALDI mass spectrometry, Talanta, 83 (2), 527-534, 2010.
4. Hui-Fen Wu, K. Suresh Kumar and Lokesh Shastri, Electrostatically self -assembled–azides on zinc sulfide nanoparticles as multifunctional nanoprobes for peptide and protein analysis in MALDI-TOF MS, Talanta, 82 (2), 540 – 547, 2010.
5. K. Suresh Kumar and Hui-Fen Wu, Interference free detection for small molecules: Probing the Mn^{2+}-doped effect and cysteine capped effect on the ZnS nanoparticles for coccidiostats and peptide analysis in SALDI-TOF MS, *Analyst*, 135 (5), 1115 – 1123, 2010.
6. K. Suresh Kumar and Hui-Fen Wu, Multifunctional ZrO2 nanoparticles and ZrO2-SiO2 nanorods for improved MALDI-MS analysis of cyclodextrins, peptides, and phosphoproteins, *Analytical and Bioanalytical Chemistry*, 396, 1115 – 1125, 2010.
7. K. Suresh Kumar and Seong Ho Kang, Microchip-based capillary electrophoresis for DNA analysis in modern biotechnology: *A Review, Separation and Purification Reviews,* 38 (3), 242–288, 2009.
8. Lokesh A. Shastri, K. Suresh Kumar, Hui-Fen Wu, Cysteine-capped ZnSe quantum dots as affinity and accelerating probes for microwave enzymatic digestion of proteins via direct matrix-assisted laser desorption/ionization time-of-flight mass spectrometric analysis, *Rapid Commun. Mass Spectrom*. 23 (15), 2247–2252, 2009.
9. K. Shrivas, K. Suresh Kumar and Hui-Fen Wu, Quantum dots laser desorption /ionization mass spectrometry: multifunctional CdSe quantum dots as the matrix, concentrating probes and acceleration for microwave enzymatic digestion for peptide analysis and high resolution detection of proteins in a linear MALDI-TOF MS, *Proteomics,* 9 (10), 2656–2667, 2009.

In: Biochemistry Researcher Biographical Sketches … ISBN: 978-1-62081-308-9
Editors: J. T. Walker and C. M. Jones

Chapter 13

Jung-Bum Lee

Affiliation:

Ph. D

Address:

2630 Sugitani, Toyama, Toyama 930-0194, Japan

Date of Birth:

Jan. 2, 1971

Education:

B.S. (1995) Department of Biochemistry, Faculty of Sciences, Saitama University, Japan
M.S. (1997)Graduate School of Sciences and Technology, Saitama University, Japan
Ph. D. (2000) Graduate School of Pharmaceutical Sciences, Toyama Medical and Pharmaceutical University, Japan

Research and Professional Experience:

2000 - Assistant Professor, Faculty of Pharmaceutical Sciences, Toyama Medical and Pharmaceutical University, Japan
2006 - Assistant Professor, Graduate School of Medicine and Pharmaceutical Sciences for Research, University of Toyama, Japan

Professional Appointments:

Assistant Professor

Honors:

2009 The JSP (The Japanese Society of Pharmacognosy) Award for Young Scientist

Publications Last 3 Years:

(1) Ogura F., Hayashi K., Lee J.-B., Kanekiyo K., & Hayashi T. "Evaluation of an Edible Blue-Green Alga, Aphanothece sacrum, for Its Inhibitory Effect on Replication of Herpes Simplex Virus Type 2 and Influenza Virus Type A", *Biosci. Biotechnol. Biochem.*, 74, 1687-1690 (2010).

(2) Lee J.-B., Ohta, Y., Hayashi, K., & Hayashi T. "Immunostimulating effects of a sulfated galactan from Codium fragile" *Carbohydr. Res.*, 345, 1452-1454 (2010).

(3) Lee J.-B., Koizumi, S., Hayashi K., & Hayashi T. "Structure of rhamnan sulfate from the green alga Monostroma nitidum and its anti-herpetic effect" Carbohydr. Polym., 81, 572-577 (2010).

(4) Dong C.X., Hayashi K., Lee J.-B., & Hayashi T. "Characterization of structures and antiviral effects of polysaccharides from Portulaca oleracea L.", *Chem. Pharm. Bull.,* 58, 507-510 (2010).

(5) Hayashi K., Hayashi T., & Lee J.-B. "Evaluation and application of natural products for viral infections", *Yakugaku Zasshi*, 130, 171-176 (2010).

(6) Li B., Lee J.-B., Hayashi K., Ito Y., Sun WJ., Wang X., Kano Y., & Hayashi T. "Two new monoterpenes from Sibiraea angustata", *J. Nat. Med.*, 64, 89-92 (2010).

(7) Ohta Y., Lee J.-B., Hayashi K., & Hayashi T. "Isolation of Sulfated Galactan from Codium fragile and Its Antiviral Effect", *Biol. Pharm. Bull.*, 32, 892-898 (2009).

(8) Tsujii K., Andoh T., Ui H., Lee J.-B., & Kuraishi Y. "Involvement of tryptase and proteinase-activated receptor-2 in spontaneous itch-associated response in mice with atopy-like dermatitis", *J. Pharmacol. Sci.*, 109, 388-395 (2009).

(9) Rattanathongkom A., Lee, J.-B., Hayashi K., Sripanidkulchai B., Kanchanapoom T., & Hayashi T. "Evaluation of Chikusetsusaponin IVâ€Ša Isolated from Alternanthera philoxeroides for Its Potency Against Viral Replication", *Planta Medica*, 75, 829-835 (2009).

(10) Tsujii K., Andoh T., Lee J.-B., & Kuraishi Y. "Activation of Proteinase-Activated Receptors Induces Itch-Associated Response through Histamine-Dependent and -Independent Pathways in Mice", *J. Pharmacol. Sci.*, 108, 385-388 (2008).

(11) Kanekiyo K., Hayashi K., Lee J.-B., Takenaka H., & Hayashi T. "Structure and Antiviral Activity of an Acidic Polysaccharide from an Edible Blue-Green Alga, Nostoc flagelliforme", *Yakugaku Zasshi*, 128, 725-731 (2008).

(12) Nakano T., Andoh T., Lee J.-B., & Kuraishi Y. "Different Dorsal Horn Neurons Responding to Histamine and Allergic Itch Stimuli", *Neuroreport,* 19, 723-726 (2008).

(13) Nakano T., Andoh T., Tayama M., Kosaka M., Lee J.-B., & Kuraishi Y. "Effects of Topical Application of Tacrolimus on Acute Itch-associated Responses in Mice", *Biol. Pharm. Bull.,* 31, 752-754 (2008).

In: Biochemistry Researcher Biographical Sketches …
Editors: J. T. Walker and C. M. Jones
ISBN: 978-1-62081-308-9

Chapter 14

C.-Y. Cynthia Lin

Affiliation:

University of California at Davis

Address:

Agricultural and Resource Economics
University of California at Davis
One Shields Avenue
Davis, CA 95616

Education:

Ph.D., Economics, Harvard University, 2006
A.M., Economics, Harvard University, 2005
A.B., Environmental Science and Public Policy, summa cum laude, Harvard University, 200

Professional Appointments:

Assistant Professor, Department of Agricultural and Resource Economics, and Department of Environmental Science and Policy, University of California at Davis, 2006 to present

Honors:

University of California at Davis Hellman Fellow (2011)

International Society for New Institutional Economics Award for the Best Ph.D. Dissertation (2006)

Stone Fellow Award for Best Paper Written by a Doctoral Student in Environmental and Resource Policy (2006)

Repsol YPF-Harvard Kennedy School Fellows Program Conference Travel Grant (2006)

The Partnership University Fellow (2005-2006)

Environmental Economics Program at Harvard University Pre-Doctoral Conference Travel Grant (2004)

Repsol YPF-Harvard Kennedy School Fellows Program Conference Travel Grant (2004)

1st Lindau Meeting of the Winners of the Bank of Sweden Prize in Economic Sciences in Memory of Alfred Nobel, young economist participant, selected to give closing remarks on behalf of students (2004)

Harvard University Department of Economics Research and Travel Award (2004)

Repsol YPF-Harvard Kennedy School Fellows Program Research Travel Grant (2004)

Harvard Committee on Undergraduate Education Certificate of Distinction in Teaching (Fall 2003)

Repsol YPF-Harvard Kennedy School Pre-Doctoral Fellowship in energy policy (2003)

Jens Aubrey Westengard Scholarship (2003)

Harvard Graduate School of Arts and Sciences Summer Research Award (2002)

Harvard Committee on Undergraduate Education Certificate of Distinction in Teaching (Fall 2002)

National Science Foundation Graduate Research Fellowship (2001)

Jacob K. Javits Fellowship (2001, declined in favor of EPA fellowship)

Environmental Protection Agency Science to Achieve Results Graduate Fellowship (2000-2003)

Rita Ricardo-Campbell Fellowship in Economics (2000)

Environmental Economics Program at Harvard University Pre-Doctoral Fellow (2000-2006)

Thomas Temple Hoopes Prize for the top 64 Harvard senior theses (2000)

Donald and Cathleen Pfister Prize for excellence in the natural sciences (2000)

Junior-year Phi Beta Kappa (1999)

Morris K. Udall Scholarship for excellence in environmental policy (1999)

Harvard College Research Program research fellowship (1999)

President Chao Nee Memorial Scholarship (1999)

John Harvard Scholarship (1997, 1998, 1999, & 2000)

Elizabeth Cary Agassiz Scholarship (1997, 1998 & 1999)

Detur Book Prize (1997)

First Place Grand Award at International Science and Engineering Fair (1996)

Publications Last 3 Years:

PEER-REVIEWED PUBLICATIONS

Ghandi, A., & Lin, C.-Y.C. (forthcoming). Do Iran's buy-back service contracts lead to optimal production?: The case of Soroosh and Nowrooz. Energy Policy.

Leighty, W., & Lin, C.-Y.C. (forthcoming). Tax policy can change the production path: A model of optimal oil extraction in Alaska. *Energy Policy*.

Lin, C.-Y.C. (2011). Learning an opponent's strategy in Cournot competition. International *Journal of Strategic Management*, 11 (1), 94-112.

Lin, C.-Y.C. (2011). Estimating supply and demand in the world oil market. *Journal of Energy and Development*, 34 (1), 1-32.

Corderi, D., & Lin, C.-Y.C. (2011). Measuring the social rate of return to R&D in coal, petroleum and nuclear manufacturing: A study of the OECD countries. *Energy Policy*, 39 (5), 2780-2785.

Pfeiffer, L., & Lin, C.-Y.C. (2010). The effect of irrigation technology on groundwater use. *Choices,* 25 (3).

Lin, C.-Y.C. (2010). A spatial econometric approach to measuring pollution externalities: An application to ozone smog. *Journal of Regional Analysis and Policy,* 40 (1), 1-19.

Adjemian, M.K., Lin, C.-Y.C., & Williams, J. (2010). Estimating spatial interdependence in automobile type choice with survey data. *Transportation Research Part A: Policy and Practice*, 44, 661-675.

Lin, C.-Y.C. (2010). How should standards be set and met?: On the allocation of regulatory power in a federal system. B.E. *Journal of Economic Analysis and Policy: Topics*, 10 (1), Article 51.

Lin, C.-Y.C. (2010). Instability, investment, disasters, and demography: Natural disasters and fertility in Italy (1820-1962) and Japan (1671-1965). *Population and Environment,* 31 (4), 255-281.

Lin, C.-Y.C., & Prince, L. (2009). The optimal gas tax for California. *Energy Policy*, 37 (12), 5173-5183.

Farinelli, B., Carter, C., Lin, C.-Y.C., & Sumner, D. (2009). Import demand for Brazilian ethanol: a cross-country analysis. *Journal of Cleaner Production* (special issue on International Trade in Biofuels), 17, S9-S17.

Lin, C.-Y.C. (2009). Estimating strategic interactions in petroleum exploration. *Energy Economics*, 31 (4), 586-594.

Lin, C.-Y.C., Meng, H., Ngai, T.Y., Oscherov, V., & Zhu, Y.H. (2009). Hotelling revisited: Oil prices and endogenous technological progress. *Natural Resources Research*, 18 (1), 29-38.

Lin, C.-Y.C. (2009). Insights from a simple Hotelling model of the world oil market. *Natural Resources Research*, 18 (1), 19-28.

Lin, C.-Y.C. (2008). An evaluation of Keynes' projected possibilities. *The American Journal of Economics and Sociology,* 67 (2), 315-329.

Lin, C.-Y.C., & Wagner, G. (2007). Steady-state growth in a Hotelling model of resource extraction. *Journal of Environmental Economics and Management*, 54 (1), 68-83.

Lin, C.-Y.C. (2005). The investment timing game in petroleum production: An econometric model. *Physica A*, 355 (1), 62-68.

Lin, C.-Y.C., Jacob, D.J., & Fiore, A.M. (2001). Trends in exceedances of the ozone air quality standard in the continental United States, 1980-1998. *Atmospheric Environment,* 35, 3217-3228.

Lin, C.-Y.C., Jacob, D.J., Munger, J.W., & Fiore, A.M. (2000). Increasing background ozone in surface air over the United States. *Geophysical Research Letters*, 27 (21), 3465-3468.

OTHER PUBLICATIONS

Lin, C.-Y.C. (forthcoming). Strategic interactions during oil exploration in the Gulf of Mexico. In *Energy Exploration*. New York: Nova Science Publishers, Inc.

Lin, C.-Y.C. (2011). An assessment of the effectiveness of California's local air pollution controls on agricultural sources. In M. Khallaf (Ed.), *Air Pollution*. InTech - Open Access Publisher.

Heres del Valle, D.R., & Lin, C.-Y.C. (2011). Sub-national state's climate policy: The case of California. In I. Galarraga, M. Gonzalez-Eguino, & A. Markandya (Eds.), *Handbook of Sustainable Energy*. United Kingdom: Edward Elgar.

Lin, C.-Y.C. (2011). Fossil fuels. *World Book Encyclopedia*. Chicago: World Book.

Lin, C.-Y.C. (2011). The effects of an E10 ethanol-blend policy on California. California State Controller John Chiang Statement of General Fund Cash Receipts and Disbursements, 5 (5), 6-7.

Lin, C.-Y.C. (2011). A Hotelling model of steady-state growth in coal extraction. In J.J. Stewart (Ed.), *Coal Extraction*. New York: Nova Science Publishers, Inc.

Lin, C.-Y.C. (2010). *Gasoline price volatility*. California State Controller John Chiang Statement of General Fund Cash Receipts and Disbursements, 4 (11), 6-8.

Lin, C.-Y.C., Zhang, W., Rouhani, O., & Prince, L. (2009). The implications of an E10 ethanol-blend policy for California. *Agricultural and Resource Economics* Update, 13 (2), 1-4.

Lin, C.-Y.C. (2009). Biofuels for transport: Global potential and implications for sustainable energy and agriculture (book review). *American Journal of Agricultural Economics,* 91(4), 1155-1157.

Pfeiffer, L., & Lin, C.-Y.C. (2009). Incentive-based groundwater conservation programs: Perverse consequences? *Agricultural and Resource Economics Update*, 12 (6), 1-4.

Lin, C.-Y.C. (2008). On the distribution of regulatory power between state and local governments. In A. Tavidze (Ed.), *Progress in Economics Research, Volume 12* (pp. 189-208). New York: Nova Science Publishers, Inc.

Lin, C.-Y.C. (2007). *California's gasoline tax.* California State Controller John Chiang Statement of General Fund Cash Receipts and Disbursements, 1 (7), 5-8.

Lin, C.-Y.C. (2007). Estimating and testing a dynamic model of OPEC and non-OPEC. In Developing and Delivering Affordable Energy in the 21st Century. 27th USAEE/IAEE North American Conference Online Proceedings.

Lin, C.-Y.C., & Leighty, W.W. (2007). Government leasing policy and the multi-stage investment timing game in offshore petroleum production. In Developing and Delivering Affordable Energy in the 21st Century. 27th USAEE/IAEE North American Conference Online Proceedings.

Lin, C.-Y.C. (2006). Environmental federalism and regulatory delegation: An incomplete contracting approach. In R. Bleischwitz & O. Budzinski (Eds.), Environmental *Economics: Institutions, Competition, Rationality* (pp. 129-149). Berlin: VWF.

Lin, C.-Y.C. (2006). Oil prices, stock effects and endogenous technological progress. In C. Frankhouse (Ed.), *Economics of Agriculture and Natural Resources* (pp. 111-125). New York: Nova Science Publishers, Inc.

Lin, C.-Y.C. (2006). Three essays on the economics of the environment, energy and externalities. Ph.D. dissertation, Harvard University.

Lin, C.-Y.C. (2005). Estimating annual and monthly supply and demand for world oil: A dry hole? In W.W. Hogan (Ed.), Repsol YPF - Harvard Kennedy School Fellows 2003-2004 Research Papers (pp. 213-249). Cambridge, MA: Harvard University.

Lin, C.-Y.C. (2005). The multi-stage investment timing game in offshore petroleum production: A framework for an econometric model. In W.W. Hogan (Ed.), Repsol YPF - Harvard Kennedy School Fellows 2003-2004 Research Papers (pp. 201-211). Cambridge, MA: Harvard University.

Lin, C.-Y.C. (2005). Optimal world oil extraction: Calibrating and simulating the Hotelling model. In W.W. Hogan (Ed.), Repsol YPF - Harvard Kennedy School Fellows 2003-2004 Research Papers (pp. 251-266). Cambridge, MA: Harvard University.

Lin, C.-Y.C. (2004). Regulatory delegation: Theory and application to environmental federalism. In G.T. Papanikos (Ed.), Competition, Regulation and Protection: Essays from an International Conference on Industrial Organization, Law and Economics (pp. 79-90). Athens: Athens Institute for Education and Research.

Lin, C.-Y.C. (2001). The scholar as activist: A colloquium explores these dual roles. *Harvard Graduate School of Arts and Sciences Bulletin*, XXX (9), 1-3.

Lin, C.-Y.C. (2000). Trends in ozone smog. Senior thesis, Harvard University.

In: Biochemistry Researcher Biographical Sketches ... ISBN: 978-1-62081-308-9
Editors: J. T. Walker and C. M. Jones

Chapter 15

Sigrid Mennickent

Affiliation:

Faculty of Pharmacy, University of Concepción, Concepción, Chile

Address:

P.O BOx 237, University of Concepción, Concepión, Chile

Date of Birth:

12-03-1962

Education:

Pharmacist

Professional Appointments:

Master in Pharmaceuticas Sciences (MSc), Associate Proffessor, Faculty of Pharmacy, University of Concepción

Honors:

Teaching prize 2007-2008.
Teaching prize 2009-2010.
Academic prize 2009-2011.

Publications Last 3 Years:

"Efectos pleiotrópicos de las estatinas". S.Mennickent, M.Bravo, C.Calvo, M.Avello. Artículo de revisión. *Rev. Med. Chile* 136, 775-782 (2008). SCIELO. Factor de impacto: 0.470

"Phospholipids bilayers as molecular models for drug-membrane interactions. The case of the antiepileptics phenytoin and carbamazepine". M.Suwalsky, S.Mennickent, F.Villena, C.P.Sotomayor. *Molecular Symposia* 269, 119-127 (2008). Corriente principal no ISI.

"Gripe aviar: situación actual". S.Mennickent. Ciencia ahora. 21, 8-12 (2008). Corriente principal no ISI.

· "Aspectos generales de la homeopatía". M.Avello, C.Avendaño, S.Mennickent. *Rev. Méd.* Chile. 137, 115-120 (2009). SCIELO. Factor de impacto: 0.470

· "Instrumental planar chromatographic method for determination of carbamazepine in human serum". Sigrid Mennickent [1*]; Ricardo Fierro [1]; Mario Vega [2]; Marta de Diego [1.], C.G. Godoy [1], *J.Sep.Sci.* 32 (9), 1454-1458 (2009). ISI. Factor de Impacto: 2.631

PAPER CITADO EN CAMAG CBS SEPTIEMBRE 2009 (PÁG. 32C).

"Extractos Antioxidantes y Antimicrobianos de Aristotelia chilensis y Ugni molinae y sus Aplicaciones como Preservantes en Productos Cosméticos". Avello, M., Valdivia, R., Sanzana, R., Mondaca, M.A., Mennickent, S., Aeschlimann, V., Bittner, M. y J. Becerra. *Boletín Latinoamericano del Caribe de Plantas Medicinales* 8(6), 479-486 (2009). ISI. Factor de impacto: 0.380

"Enfermedad celíaca: "enfermedad emergente en Chile?". S.Mennickent. Ciencia ahora, 23: 1-5 (2009). Corriente principal no ISI.

"Los probióticos y su utilidad terapéutica". S.Mennickent, K.Green, Ciencia ahora 24: 31-38 (2009). Corriente principal no ISI.

"Validated instrumental planar chromatographic method for quantification of fluphenazine hydrochoride in injections". Sigrid Mennickent [1*]; Jonathan Contreras[1]; Cristóbal Reyes[1]; Mario Vega [2]; Marta de Diego [1]. JPC, 23, 1-13 (2010). ISI. Factor de impacto: 1.247

"Determination of clozapine in serum of patients with schizophrenia as a measurement of medication compliance".S.Mennickent [a*], A.Sobarzo [a], M.Vega [b], M.de Diego [a], G.Godoy [a,] P.Rioseco[c], L. Saavedra[c]. International Journal of Psychiatry in Clinical Practice,14, 41-46 (2010). ISI. Factor de impacto: 0.490

"Structural effects of verapamil on cell membranes and molecular models". M. Suwalsky, M. Muñoz, S. Mennickent, C.P. Sotomayor, S. Bolognin. J.Chil. Chem. Soc. 55(1),1-4 (2010). ISI. Factor de impacto: 0.629.

"Stress degradation studies of ramipril by a validated stability-indicating liquid chromatographic method". M. de Diego, C.G.Godoy, S.Mennickent, M.Olivares, R.Godoy. J.Chil.Chem.Soc. 55(4), 450-453 (2010). ISI. Factor de impacto: 0.629.

"Quantitative determination of fluoxetine in human serum by high performance thin layer chromatography". S.Mennickent, R.Fierro, M.Vega, M.de Diego, C.G.Godoy. J.Sep.Sci. 33, 2206-2210 (2010). ISI. Factor de Impacto: 2.631

· "Chemical stability of ceftriazone by a validated stability-indicating liquid chromatographic method". M.de Diego, C.G.Godoy, S.Mennickent. J.Chil.Chem.Soc. 55 (3): 335-337 (2010). ISI. Factor de impacto: 0.629.

"Chemical stability of enalapril maleate drug substance and tablets by a stability-indicating liquid chromatographic method". M.de Diego, R.Godoy, C.G.Godoy, S.Mennickent. Quimica Nova 34(3): 450-454 (2011). ISI. Factor de impacto: 0.898.

"Quantification of lamotrigine in human serum by high performance thin layer chromatograhy". Mennickent S, Fierro R, Vega M, de Diego M, Godoy C. JPC 24(3): 222-226 (2011). ISI. Factor de impacto: 1.247.

"Effects of phenylpropanollamine (PPA) on in vitro human erythrocyte membranes and molecular models". Suwalsky M, Zambrano P, Mennickent S, Villena F, Sotomayor C.P. *Biochemical and Biophysical Research Communication* 406:320-325 (2011). ISI. Factor de impacto: 2.595.

"A validated stability-indicating LC method for simultaneous determination of enalapril and hydrochlorothiazide in pharmaceutical samples". Marta De Diego*, Sigrid Mennickent, Gloria Godoy, Valeska Miranda. *Current Pharmaceutical Analysis* 7 (4): 248-252 (2011). ISI. Factor de impacto: 1.710.

"Quantitative determination of propranolol in human serum by high performance thin layer chromatography". S.Mennickent, M.Vega M., M.Vega H, M.de Diego, R.Fierro. JPC. Aceptado Julio 2011. ISI. Factor de impacto: 1.247.

"High performance thin layer chromatographic determination of nifedipine in human serum after liquid-liquid extraction". S.Mennickent, J.Contreras, B.Schulz, M.de Diego, M.Vega. Química Nova. Aceptado Agosto 2011. ISI. Factor de impacto: 0.898.

In: Biochemistry Researcher Biographical Sketches …
Editors: J. T. Walker and C. M. Jones
ISBN: 978-1-62081-308-9

Chapter 16

Igho Onakpoya

Affiliation:

University of Oxford, United Kingdom

Address:

Dept Of Primary Care Health Sciences, University Of Oxford, 23-38 Hythe Bridge Street, Oxford, United Kingdom. Ox1 2et

Date of Birth:

09-JAN, 1974

Education:

MBBS (Benin, Nigeria); MSc (Aberdeen, Uk)

Research and Professional Experience:

Assessing the methodology of clinical trials, efficacy and safety of therapeutic interventions, anti-obesity therapies, adverse drug reactions, as well as drug-drug interactions.

Professional Appointments: Research Associate in Pharmacovigilance; GSK Research Fellow in Obesity

Publications Last 3 Years:

Onakpoya I, Ernst E. Dietary supplements for weight loss. In: Kanarek RB, Lieberman HR (eds) *Diet, Brain, Behavior: Practical Implications*. 2012. CRC Press, Florida, pp 153-167.

Onakpoya IJ, Posadzki PPP, Watson LK, Davies LA, Ernst E. The efficacy of long-term conjugated linoleic acid (CLA) supplementation on body composition in overweight and obese individuals: a systematic review and meta-analysis of randomized clinical trials. *Eur J Nutr* 2011 Oct 12 [Epub ahead of print].

Onakpoya IJ, Perry R, Zhang J, Ernst E. Efficacy of calcium supplementation for management of overweight and obesity: systematic review of randomized clinical trials. *Nutr Rev* 2011; 69(6): 335343.

Onakpoya I, Aldaas S, Terry R, Ernst E. The efficacy of Phaseolus vulgaris as a weight-loss supplement: a systematic review and meta-analysis of randomised clinical trials. *BJN* (2011); 106(02): 196-202.

Onakpoya IJ, Wider B, Pittler MH, Ernst E. Food supplements for body weight reduction: a systematic review of systematic reviews. *Obesity* (2011); 19(2): 239-244.

Onakpoya I, Davies L Ernst E. Efficacy of herbal supplements containing Citrus aurantium and synephrine alkaloids for the management of overweight and obesity: a systematic review. *FACT* 2011; 16(4): 254-260.

Clement YN, Onakpoya I, Hung SK, Ernst E. Effects of herbal and dietary supplements on cognition in menopause: A systematic review. *Maturitas.* 2011; 68 (3): 256-263.

Onakpoya I, Terry R, Ernst E. The Use of Green Coffee Extract as a Weight Loss Supplement: A Systematic Review and Meta-Analysis of Randomised Clinical Trials. *Gastroenterol Res Pract* Vol 2011, Article ID 382852. doi:10.1155/2011/382852.

Onakpoya I, Hung SK, Perry R, Wider B, Ernst E. The Use of Garcinia Extract (Hydroxycitric acid) as a weight loss supplement: a systematic review and meta-analysis of randomized clinical trials. *J Obes,* Vol 2011, no.3, Article ID 509038, 9 pages. doi: 10.1155/2011/509038.

Onakpoya I, Ho S. Is Traditional Chinese Medicine Effective for the management of Psoriasis? *FACT 2011*; 16(2): 143-144.

Onakpoya I. Supplement containing Camellia sinensis (green tea) may help treat obesity. *FACT 2010*; 15(1): 15-16.

In: Biochemistry Researcher Biographical Sketches …
Editors: J. T. Walker and C. M. Jones
ISBN: 978-1-62081-308-9

Chapter 17

Hugo H. Ortega

Affiliation:

Associate Professor of Cellular Biology
Research Career Member of the National Scientific Research Council (CONICET, Argentina).

Date of Birth:

16, 1971

Education:

National University of Litoral (Argentina)	Ph.D.	2002	Biological Sciences
National University of Litoral (Argentina)	DMV	1997	Veterinary Sciences

A. Positions and Honors. Positions and Employment:

1992-1997 Tutor, Department of Histology and Embryology, Faculty of Veterinary Sciences, National University of
Litoral, Argentina.

1997-2000 Instructor and Fellow, Department of Cellular Biology, Faculty of Veterinary Sciences, National University of Litoral, Argentina.

1997-2002 Doctoral Fellow of the National Scientific Research Council (CONICET, Argentina), Laboratory of Endocrinology and Endocrine-dependent tumors. Department of Human Physiology, Faculty of Biochemistry and Biological Sciences, UNL, Santa Fe, Argentina.

1998-present Visiting Professor, Department of Theriogenology, Faculty of Veterinary Sciences, National University of Litoral, Argentina.

1999 Research Fellow at the Department of Histology and Embryology, Institute of Biomedical Science, University of São Paulo, Brazil.

2000-2010 Professor, Department of Cellular Biology, Faculty of Veterinary Sciences, National University of Litoral, Argentina.

2004-2005 Posdoctoral Fellow of the National Scientific Research Council (CONICET, Argentina).

2005-present Research Career Member of the National Scientific Research Council (CONICET, Argentina).

2010-2010 Associated Professor, Department of Cellular Biology, Faculty of Veterinary Sciences, National

University of Litoral, Argentina.

Other Experience and Professional Memberships:

1997-present Director, Center of Experimental Biology and Biotery, Faculty of Veterinary Sciences, National University

of Litoral, Argentina.

1995-present Member, Society of Morphological Sciences, Argentina.

1997-present Member, Society of Specialist in Laboratory Animals, Argentina.

2002-present Member, Society of Veterinary Pathology, Argentina.

2002-present Reviewer of grants, National Agency of Technological and Scientific Promotion, Argentina.

2002-present Reviewer of grants, National University of Litoral, Argentina

2003-present Reviewer of International Cooperation Programs, National University of Litoral

2005-present Reviewer of journals, Medical Science Monitor; Reproduction in Domestic Animals; Brain Research

Bullerin; Theriogenology, Histology and Histopathology, Reproduction Fertility and Developmen; International Journal of Biomedical Science; Research in Veterinary Sciences; Journal of Molecular Endocrinology, Reproduction, etc

2005-present Reviewer of Research Career members, National Scientific Research Council (CONICET, Argentina).

2006-present Reviewer of grants, beroamerican Program of Science and Technology for Development (CYTED),

2009-present Member, Society for the Study of Reproduction, USA.

Honors:

2002 National Award to Excellence, National Institute for Excellence, Argentina

2007 Ten Outstanding Young People of the Province, Worldwide Federation of Young Leaders and Entrepreneurs

2009 Invited Researcher - Molecular and Integrative Physiology Department - University of Michigan

2010 Young Researcher Award – National Academy of Natural Sciences – Argentina

B. Selected peer-reviewed publications (in chronological order). (Publications selected from 80 peer-reviewed publications)

1. Salvetti, N.; Ricci, N.; Dallard, B.; Lorente J.A; Iguzquiza I.; Ortega, H.H. (2000). "Lectinhistochemical and Cytometrical Evaluation of Rats Corpus Luteum at the end of Pregnancy". *Anatomia Histologia Enbryologia* (Journal of Veteterinary Medicine. Series C). 29:129-34.
2. Kass, L.; Varayoud, J.; Ortega, H.H.; Muñoz de Toro, M.; Luque, E.H. (2000). "Detection of bromodeoxyuridine in formalin-fixed tissue. DNA denaturation following microwave or enzymatic digestion pretreatment is required". *European Journal of Histochemistry.* 44:185-191.
3. Maffini, M.V.; Ortega, H.H.; Stocker, C.; Giardina, R.H.; Luque E.H.; Muñoz de Toro, M.M. (2001). "Fine-needle aspiration biopsies from early stage prostate carcinoma: bcl-2 expression and its association with tumor ploidy and nuclear morphometry". *Pathology Research and Practice*. 197:487-492.
4. Kass, L.; Ramos, J.G.; Ortega, H.H.; Montes, G.S.; Bussmann, L.; Luque, E.H.; Muñoz de Toro, M. (2001). "Relaxin has a minor role in rat mammary gland growth and differentiation during pregnancy". *Endocrine* 15(3):263-269.
5. Ortega, H.H.; Joazeiro, P.; Muñoz de Toro, M.M.; Luque, E.H.; Montes,G.S.M. (2001). "Differential distribution of fibres of the collagenous and elastic systems and glycosaminoglycans in the rat pubic joint". *Journal of Submicroscopic Cytology and Pathology* 33(3):463-472.
6. Vicario, J.; Piva, J.R.; Pierini, A.; Campos C.; Canal A.M.; Ortega, H.H.; Pfeiffer, H.; Gerardo, L.; Fendrich, I.; Novero R. (2002). "Transcoronary Sinus Delivery of Autologous Bone Marrow an Angiogenesis in Pig Models with Myocardial Injury". *Cardiovascular Radiation Medicine*. 3(2): 91-94.
7. Ortega, H.H.; Muñoz de Toro, M.M. ; Luque E.H.; Montes G.S. (2003). "Morphological Characteristic of the Interpubic Joint (Symphysis pubica) of Rat, Guinea Pig and Mice in Different Physiological situation. A Comparative Study.". Cell Tissues Organs. 173(2):105-114.
8. Salvetti, N.; Gimeno, E.J.; Canal, A.M.; Lorente, J.A., Ortega, H.H. (2003) "Histochemical study of the extracellular matrix components in the follicular wall of induced polycystic ovaries". *Brazilian Journal of Morphological Sciences*. 20(1):93-100.
9. Rodriguez, H.; Kass, L.; Varayoud, J.; Ramos J.G.; Ortega, H.H.; Muñoz de Toro, M.; Luque, H. (2003) "Collagen remodelling in the guinea pig uterine cervix at term is associated with a decrease in PR and an increase in ER expressions". *Mollecular Human Reproduction.* 9(12):807-813-
10. Rodriguez, H.A.; Ortega, H.H.; Ramos, J.G.; Munoz-De-Toro, M.M.; Luque, E.H. (2003) "Guinea pig interpubic joint (symphysis pubica) relaxation at parturition: Underlying cellular processes that resemble an inflammatory response". *Reproductive Biology and Endocrinology* 1:113.
11. Ortega, H.H.; Lorente, J.A.; Salvetti, N.R. (2004) "Immunohistochemical study of intermediate filament and neuroendocrin markers expression on Leydig cells of laboratory rodents." *Anatomia Histologia Enbryologia* (Journal of Veterinary Medicine - Series C). 33 (5): 309-315.

12. Salvetti, N.R.; Gimeno, E.J.; Lorente, J.A.; Ortega, H.H. (2004) "Expression of cytoskeletal proteins in the follicular wall of induced ovarian cysts". *Cell Tissues Organs.* 178(2):117-125.
13. Ortega, H.H.; Lorente, J.A.; Mira, G.A.; Baravalle, C.; Salvetti, N.R. (2004) "Constant light exposure cause dissociation in gonadotrophins secretion and inhibits partially neuroendocrine differentiation of Leydig cells in adult rats". *Reproduction in Domestic Animals.* 39(6):417-423.
14. Salvetti, N.; Canal, A.M.; Gimeno, E.J.; Ortega, H.H. (2004) "Polycystic Ovarian sindrome: temporal characterization of the induction and reversion process in an experimental model". *Brazilian Journal of Veterinary Research and Animal Science.* 41:389-395.
15. Dallard, B.E.; Ortega, H.H.; Lorente, J.A.; Romano, G.S. (2005) "Immunolocalization and expression of insulin-like growth factor I (IGF-I) in the ovine mammary gland during mammogenesis, lactation and involution". *Small Rumiant Research.* 58(1):1-11.
16. Ortega, H.H.; Salvetti, N.R.; Baravalle, C.; Lorente, J.A.; Mira, G.A. (2006) "Oestradiol Induced Inhibition of neuroendocrine Markers Expression in Leydig cells of adult rats." *Reproduction in Domestic Animals.* 41(3):204-209.
17. Baravalle, C.; Salvetti, N.R.; Mira, G.A.; Pezzone, N.; Ortega, H.H. (2006) "Microscopic Characterization of Follicular Structures in Letrozole -Induced Polycystic Ovarian Syndrome in the Rat". *Archives of Medical Research.* 37(7):830- 839.
18. Baravalle, C.; Salvetti, N.R.; Mira,G.A.; Lorente, J.A.; Ortega, H.H. (2007) "Role of ACTH in the pathogenesis of Polycystic Ovarian Syndrome in the rat: hormonal profiles and ovarian morphology". *Physiological Research.* 56(1):67-78.
19. Ortega, H.H.; Salvetti, N.R.; Amable, P.; Dallard, B.E.; Baravalle, C.; Barbeito, C.G.; Gimeno, E.J. (2007) "Intraovarian localization of growth factors in induced cystic ovaries in rats". *Anatomia Histologia Enbryologia* (Journal of Veterinary Medicine - Series C). 36(2):94-102.
20. Salvetti, N.R.; Acosta, J.C.; Gimeno, J.E.; Muller, L.A.; Mazzini, R.A.; Taboada, A.F.; Ortega, H.H. (2007) "Estrogen Receptors a and 13 and Progesterone Receptors in Ovarian Follicles of Cows with Cystic Ovarian Disease". *Veterinary Pathology.* 44(3):373-378.
21. Ortega, H.H.; Salvetti, N.R.; Müller, L.A.; Amable, P.; Lorente, J.A.; Barbeito, C.G.; Gimeno, E.J. (2007) "Characterization of cytoskeletal proteins in follicular structures of cows with Cystic Ovarian Disease". *Journal of Comparative Pathology.* 136(4):222-230.
22. Palomar, M.M.; Acosta, J.C.; Salvetti, N.R.; Barberis, F.; Beldomenico, P.M.; Garnero, O.; Ortega, H.H. "Treatment of Cystic Ovarian Disease with Naloxone in High Production Dairy Cows". *Journal of Veterinary Pharmacology and Therapeutics.* 31(2):184-186.
23. Ortega, H.H.; Palomar, M.M.; Acosta, J.C.; Salvetti, N.R.; Dallard, B.E.; Lorente, J.A.; Barbeito, C.G.; Gimeno, E.J. (2008) "Insulin-like growth factor I in ovarian follicles and follicular fluid of cows with spontaneous and induced cystic ovarian disease". *Research in Veterinary Science.* 84(3):419-427.
24. Dallard, B.E.; Baravalle, C.; Ortega, H.H.; Ruffino, V.; Heffel S.; Calvinho, L.F. (2008) "Effect of a biological response modifier on cellular death mechanisms at drying off". *Journal of Dairy Research.* 75(2):167-175.

25. Rodriguez, H.A.; Ramos, J.G.; Ortega, H.H.; Munoz-De-Toro, M.M.; Luque, E.H. (2008) "Regional changes in the spatio-temporal pattern of progesterone receptor (PR) expression in the guinea-pig genital tract as parturition approaches". *Journal of Steroid Biochemistry and Molecular Biology*. 111:247-254.
26. Francou, M.; Durdos, M.; Salvetti, N.R.; Baravalle, C.; Rey, F.; Ortega, H.H. (2008) "Characterization of pituitary cell populations in rats with induced cystic ovarian disease". *Cell Tissues Organs*. 188(3):310-319.
27. Salvetti, N.R.; Baravalle, C.; Mira, G.A.; Gimeno E.J.; Dallard B.E.; Rey, F.; Ortega, H.H. (2009) "Heat shock protein 70 and sex steroid receptors in the follicular structures of induced ovarian cysts." *Reproduction in Domestic Animals*. 44:805-814.
28. Ortega, H.H.; Salvetti, N.R.; Padmanabhan, V. (2009) "Developmental programming: prenatal androgen excess disrupt ovarian steroid receptor balance". *Reproduction*. 137(5):865-877. (ISSN 1470-1626).
29. Salvetti, N.R.; Panzani, C.G.; Gimeno, E.J.; Neme, L.G.; Alfaro, N.S. Ortega, H.H. (2009) "An imbalance between apoptosis and proliferation contributes to follicular persistence in polycystic ovaries in rats." *Reproductive Biology and Endocrinology* 7:68.
30. Zurvarra, F.M.; Salvetti, N.R.; Mason, J.I.; Velazquez, M.M.L.; Alfaro, N.S.; Ortega, H.H. (2009) "Disruption in the Expression and Immunolocalization of Steroid Receptors and Steroidogenic Enzymes in Letrozole-induced Polycystic Ovaries in Rat". *Journal of Reproduction Fertility and Development*. 21(7):827-839.
31. Dallard, B.E.; Baravalle, C.; Ortega, H.H.; Tumini, M.; Canavesio, C.V.; Neder, V.E.; Calvinho, L.F. (2009) "Effect of a biological response modifier on expression of CD14 receptor and tumor necrosis factor-alpha in Staphylococcus aureus infected mammary glands at drying off". *Veterinary Immunology and Immunopathology*. 132(2-4):237-242.
32. Alzamendi, A.; Castrogiovanni, D.; Ortega, H.H.; Gaillard, R.C.; Giovambattista, A.; Spinedi, E. (2010) "Fructose rich diet-induced allostasis aggravates metabolic syndrome development in the Polycystic Ovary Syndrome Rat Phenotype" *Obesity*. 18(3): 441-448.
33. Ortega, H.H.; Rey, F.; Velazquez, M.M.L.; Padmanabhan, V. (2010) "Effect of prenatal steroid excess on intraovarian components of insulin signaling pathway and repated proteins". *Biology of Reproduction*. 82(6):1065-1075.
34. Dallard, B.E.; Ortega, H.H.; Iguzquiza, I.A.; Salvetti, N.R.; Quaino, O.A.; Calvinho, L.F. (2010) "The effect of a single intramammary infusion of a biological response modifier in cows at drying off". *Veterinary Research Communications*. 34(6):519-532.
35. Rosciani, A.S.; Merlo, W.A.; Insfrán, R.M.; Benítez, J.S.; Ortega, H.H. (2010) "Rol de las células mioepiteliales en el desarrollo de tumores complejos y mixtos de la glándula mamaria canina". *Revista Veterinaria*. 21(1):19-22.
36. Salvetti, N.R.; Stangaferro, M.L.; Palmar, M.M.; Alfaro, N.A.; Rey, F.; Gimeno, E.J.; Ortega, H.H. (2010) "Cell proliferation and survival mechanisms underlying the abnormal persistence of follicular cysts in bovines with cystic ovarian disease induced by ACTH" *Animal Reproduction Sciences* 122(1-2):98-110.
37. Baravalle, C.; Dallard, B.E.; Ortega, H.H.; Neder, V.E.; Canavesio, L.F.; Calvinho, L.F. (2010) "Effect of Panax ginseng on cytokine expression in bovine mammary glands at drying off." *Veterinary Immunology and Immunopathology*. 138(3):224-230.

38. Amweg, A.N.; Paredes, A.H.; Salvetti, N.R.; Lara, H.E.; Ortega, H.H. (2011) "Expression of melanocortin receptors mRNA, and direct effects of ACTH on steroids secretion in the bovine ovary." *Theriogenology*. 75 (4): 628-637.
39. Dallard, B.E.; Baravalle, C.; Andreotti, C.; Ortega, H.H.; Neder, V.; Calvinho, L.F. (2011) "Intramammary inoculation of Panax ginseng extract in cows at drying off enhances early mammary involution". *Journal of Dairy Research*. 78(1):63-71.

C. Research Support:

Ongoing Research Support

FONCYT - PICT 2005 - 38101 Ortega (PI) 03/01/07-03/01/10
National Agency of Technological and Scientific Promotion, Argentina.
Bovine cystic ovarian disease: Hormonal receptors and HSPs role in the cystogenesis.
CH/08/04 Ortega and Paredes (PI) 01/01/09-12/31/11
International Cooperative agreement supported by SECyT (Argentina) and CONICYT (Chile)
Effect of the stress on the expression of glucocorticoid receptors and the synthesis of steroid hormones in the ovary of bovine: participation in the pathogenia of the ovarian cystic diseasse.
FONCYT - PICT 2007 - 01193 Ortega (PI) 03/31/09-03/31/12
National Agency of Technological and Scientific Promotion, Argentina.
Expression and activity of elements of stress response in then ovary and their implications in the pathogenia of the cystic ovarian disease and other anestrous in cattle
FONCYT - PICT 2008 - 1952 Ortega (PI) 03/01/10-03/01/13
National Agency of Technological and Scientific Promotion, Argentina.
Involvement of Transforming Growth Factor (TGF-13) superfamily in ovarian reproductive disorders of bovines

Completed Research Support

CAID 2002 26/176 Ortega (PI) 01/01/05-12/31/05
National University of Litoral
Polycystic Ovary: Immunohistochemical characterization of ovarian alterations in a rat experimental model.
FONCYT - PICTO 2003 - 08-13243 Ortega (PI) 03/01/04-03/01/06
National Agency of Technological and Scientific Promotion, Argentina.
Comparative study of two experimental models of bovine cystic ovarian disease (COD) and their relation with growth factors and binding proteins expression.
FONCYT - PICT 2003 - 08-14734 Ortega (PI) 03/01/04-03/01/07
National Agency of Technological and Scientific Promotion, Argentina.
Bobine cystic ovarian disease: Modifications in the hormonal receptor expresion and their relation with cellular proliferation and apoptosis
2005-7-162 - CH/PA05-BXII/002 Ortega and Lara (PI) 01/01/06-12/31/07

International Cooperative agreement supported by SECyT (Argentina) and CONICYT (Chile)
Participation of neuroendocrine system in the pathogenia of Bobine cystic ovarian disease. Role: PI

CAID 2006 18/106 Ortega (PI) 01/01/06-12/31/08
National University of Litoral
Participation of the Pituitary Gland in the Pathogenia of Cystic Ovarian Disease

In: Biochemistry Researcher Biographical Sketches …
Editors: J. T. Walker and C. M. Jones
ISBN: 978-1-62081-308-9

Chapter 18

Paolo Polidori

Affiliation:

School of Pharmacy, University of Camerino

Address:

School of Pharmacy, University of Camerino, Via Circonvallazione 93, 62024 Matelica (MC), Italy

Date of Birth:

September, 30, 1963

Education:

M..Sc. in Science of Agriculture, Ph.D in Animal Husbandry

Research and Professional Experience:

1995-1996 - visitng scientist at the Muscle Biology Laboratory, University of Wisconsin, Madison, WI

1998-1999 - visiting scientist at te Department of Food Science and Technology, University of Tennessee, Knoxville, TN

Professional Appointments:

1993-2000 - Researcher at the Faculty of Veterinary Medicine, University of Camerino, Italy
2000-2005 - Associated professor of Animal Husbandry at the Faculty of Veterinary Medicine, University of Camerino, Italy
since 2006 - Full professor of Animal Nutrition at the School of Pharmacy, University of camerino, Italy

Honors:

2008-2009 - Member of the Editorial Board of the journal Small Ruminant Research, Elsevier Publisher, The Netherlands
since 2011 - Member of the Editorial Board of the Open Journal of Animal Science

Publications Last 3 Years:

1. Meat fatty acid composition of llama (Lama glama) reared in the Andean highlands. Polidori P., Renieri C., Antonini M., Passamonti P., Pucciarelli F. *Meat Science*, 2007, 75, 366-368.
2. Tenderness evaluation and mineral levels of llama (Lama glama) and alpaca (Lama pacos) meat. Polidori P., Antonini M., Torres D., Beghelli D., Renieri C. *Meat Science*, 2007, 77, 599-601.
3. Donkeyâ€™s milk protein fractions characterization.Vincenzetti S., Polidori P., Mariani P., Cammertoni N., Fantuz F., Vita A. *Food Chemistry*, 2008, 106, 640-649.
4. Protein fraction characterization of sheep milk from the Comisana breed. Vincenzetti S., Polidori P., Mariani P., Vita A. *Veterinary Research Communications*, 2008, 32 (Suppl. 1), S179-S181.
5. Quality of donkey meat and carcass characteristics. Polidori P., Vincenzetti S., Cavallucci C., Beghelli D. *Meat Science*, 2008, 80, 1222-1224.
6. Physical and chemical Characteristics of donkey meat from Martina Franca breed. Polidori P., Cavallucci C., Beghelli D., Vincenzetti S. *Meat Science*, 2009, 82, 469-471.
7. Differences of Protein Fractions Among Fresh, Frozen and Powdered Donkey Milk. Polidori P., Vincenzetti S. Recent Patents on Food, *Nutrition & Agriculture*, 2010, 2, 56-60.

In: Biochemistry Researcher Biographical Sketches ... ISBN: 978-1-62081-308-9
Editors: J. T. Walker and C. M. Jones

Chapter 19

Marzia Salmaso

Affiliation:
University of Padua (Italy)

Address:
Lab. of Genetics and Molecular Biology
University of Padua
Viale Università, 16, Agripolis Legnaro (Padova) Italy
E-mail: marzia.salmaso@gmail.com

Date of Birth: 18 of August 1973

Education:

1999, University Degree in Biology. University of Padua (Italy)
2004, Ph.D. Istituto Agrario di San Michele all'Adige (TN-Italy) -Edmund Mach Foundation and University of Padua (Italy)

Research and Professional Experience:

Research Associate 2004-
University of Padua (Italy)
- Laboratory research activities (genetics and molecular biology):
- Logistics and organization of laboratory work
- Tutoring of graduate and undergraduate students
- Didactic activities (genetics)

Research Associate 2000 and Visiting Scientist 2004

Department of Biology - Indiana University (Bloomington, USA)

Main activities and responsibilities: Evolutionary Genetics laboratory research aimed to study the origin and evolution of sunflower species.

Post-Graduate Fellowship 2001-2004 in Biology and Molecular Genetics at the Istituto Agrario di San Michele all'Adige (TN-Italy) - Edmund Mach Foundation

Post-Graduate Training 1999 Department of Biology - University of Padua (Italy)

Professional Appointments:

2004- Research Scientist University of Padua (Italy)

Publications Last 3 Years:

John M. Burke, Zhao Lai, Marzia Salmaso, Takuya Nakazato, Shunxue Tang, Adam Heesacker, Steven J. Knapp, and Loren H. Rieseberg: Comparative Mapping and Rapid Karyotypic Evolution in the Genus *Helianthus. Genetics* 2004. 167(1):449-457.

Marzia Salmaso & Loren H. Rieseberg: Crop Plant Promiscuity. *TRENDS in Ecology and Evolution* 2004. 19:293-294.

Zhao Lai, Takuya Nakazato, Marzia Salmaso, John M. Burke, Shunxue Tang, Steven J. Knapp, Loren H. Rieseberg: Extensive Chromosomal Repatterning and the Evolution of Sterility Barriers in Hybrid Sunflower Species. *Genetics* 2005. 171:291-303.

Marzia Salmaso, Giorgia Faes, Cinzia Segala, Marco Stefanini, Ilkhom Salakhutdinov, Eva Zyprian, Reinhard Toepfer, M. Stella Grando, Riccardo Velasco: Genome diversity and gene haplotypes in grapevine (*Vitis vinifera* L.), as revealed by single nucleotide polymorphisms. *Molecular Breeding* 2005. 14(4):385-395.

Marzia Salmaso, Giulia Malacarne, Michela Troggio, Giorgia Faes, Marco Stefanini, M. Stella Grando, Riccardo Velasco: A grapevine (*Vitis vinifera* L.) genetic map integrating the position of 139 expressed genes. TAG 2008. 116(8):1129-1143.

Marzia Salmaso, Renzo Dalla Valle, Margherita Lucchin: Genepool variation and phylogenetic relationships of an indigenous north-east Italian grapevine collection revealed by nuclear and chloroplast SSRs. *Genome* 2008. 51:838-855.

In: Biochemistry Researcher Biographical Sketches … ISBN: 978-1-62081-308-9
Editors: J. T. Walker and C. M. Jones

Chapter 20

Luãs Adriano Santos Do Nascimento

Affiliation: Instituto Federal do ParÃ¡-Castanhal

Address:

Psg. Doutel, 308

Date of Birth: 03/03/1981

Education: Dr

Research and Professional Experience:

Researches about catalysis, biodiesel

Professional Appointments: Prof. Dr

Publications Last 3 Years: 3

In: Biochemistry Researcher Biographical Sketches ...
Editors: J. T. Walker and C. M. Jones
ISBN: 978-1-62081-308-9

Chapter 21

Elena Timofeeva

Affiliation: Laval University

Address:

Faculty of Medicine, Department of Psychiatry and Neuroscience
Centre de recherche de l'Institut universitaire de cardiologie et de pneumologie de Québec (CRIUCPQ)
Pavillon Marguerite d'Youville, local Y3150
2725, chemin Sainte Foy
Québec (Québec), G1V 4G5
Elena.Timofeeva@fmed.ulaval.ca
Tel office 418-656-8711 ex 3749

Date of Birth: 2/04/1962

Education:

Ph.D. Laval University, Quebec, Canada

Professional Appointments:

2011 – present Associate professor
2005- 2011 Assistant professor
1992-1995 Senior Investigator
1988-1992 Investigator
1986-1988 Junior Investigator
1985-1986 Project manager

Publications Last 3 Years:

Book chapters:

Richard D and Timofeeva E (2010) Energy balance regulation: complex interplay between the autonomic and cognitive limbic brains to control food intake and thermogenesis. In: *Obesity Prevention: The Role of Brain and Society on Individual Behavior*. Dubé L, Bechara A, Dagher A, Drewnowski A, LeBel J, James P, Richard D, Yada RY (Eds). London: Elsevier Science, pp. 299-309.

Richard D and Timofeeva E (2009) Energy Homeostasis: Central integrators: PVN system. *Encyclopedia of Neuroscience*. Ed: Larry R. Squire, Elsevier, Berlin. p. 1035 – 1041.

Articles:

Baraboi ED, Timofeeva E, St-Pierre D, and Richard D (2011) Brain activation following peripheral administration of exendin-4: comparisons with eating and effects of vagotomy. *Am J Physiol, in press*.

Mitra A, Lenglos C, Martin J, Mbende N, Gagné A, and Timofeeva E (2011) Sucrose modifies c-fos mRNA expression in the brain of rats maintained on feeding schedules. *Neuroscience, 192:* 459-474.

Roy C, Roy MC, Gauvreau D, Poulin AM, Tom FQ, Timofeeva E, Richard D, and Katherine Cianflone (2011) Acute injection of ASP in the third ventricle inhibits food intake and locomotor activity in rats. *Am J Physiol, 301 (1):* E232-241.

Timofeeva E (2010) Yin and yang in regulation of alcohol and sucrose intake by CRF1 and CRF2 receptors. *Can J Addiction Med, 1 (3):* 4-11.

Martin J and Timofeeva E (2010) Intermittent access to sucrose increases sucrose licking activity and attenuates stress-induced activation of the lateral septum. *Am J Physiol, 298:* R1383-R1398.

Timofeeva E and Martin J (2009) Addiction to palatable food: neuronal and behavioral response to intermittent access to sucrose. *Can J Addiction Med, 1*: 24.

Timofeeva E, Baraboi ED, Poulin AM, and Richard D (2009) Palatable high-energy diet decreases the expression of cannabinoid type 1 receptor (CB1R) mRNA in specific brain regions in rat. *J Neuroendocrinology, 21:* 982-992.

Richard D, Guesdon B, and Timofeeva E (2009) The brain endocannabinoid system in the regulation of energy balance. *Best Practice & Research Clinical Endocrinology & Metabolism, 23*: 17-32.

In: Biochemistry Researcher Biographical Sketches …
Editors: J. T. Walker and C. M. Jones
ISBN: 978-1-62081-308-9

Chapter 22

Ana Villares

Affiliation: Instituto Nacional de Investigación y Tecnología Agraria y Alimentaria (INIA)

Address:

Centro para la Calidad de los Alimentos. Campus Universitario Duques de Soria, c/ José Tudela s/n, 42004 Soria (Spain)

Date of Birth: 15th August 1980

Education: Chemistry degree

Research and Professional Experience:

Food research in terms of food composition and beneficial properties.

Professional Appointments:

EUROANALYSIS, Innsbruck (Austria)
10th International Nutrition and Diagnostics Conference, Prague (Czech Republic)
28th International Symposium on Chromatography, Valencia (Spain)
Effost Annual Meeting, Dublin (Ireland)

Honors:

European doctorate

Publications Last 3 Years:

Palacios I, Lozano M, Moro C, D'Arrigo M, Rostagno M A, Martínez J A, García-Lafuente A, Guillamón E & Villares A (2011). "Antioxidant properties of phenolic compounds occurring in edible mushrooms". *Food Chemistry, 128*, 674-8.

Villares A, Rostagno M A, Garcia-Lafuente A, Guillamon E & Martiez J A (2011). "Content and profile of isoflavones in soy-based foods as a function of the production process". *Food and Bioprocess Technology, 4*, 27-38.

Rostagno M A, Manchon N, D'Arrigo M, Guillamon E, Villares A, Garcia-Lafuente A, Ramos A & Martinez J A (2011). "Fast and simultaneous determination of phenolic compounds and caffeine in teas, mate, instant coffee, soft drink and energetic drink by high-performance liquid chromatography using a fused-core column". *Analytica Chimica Acta, 685*, 204-11.

García - Lafuente A, Moro C, Villares A, Guillamón E, Rostagno M A, D'Arrigo M & Martínez J A (2010). "Mushrooms as a source of anti-inflammatory agents". *Anti-inflammatory and Anti-allergy Agents in Medicinal Chemistry, 9*, 125-41.

Guillamón E, García-Lafuente A, Lozano M, D'Arrigo M, Rostagno M A, Villares A & Martínez J A (2010). "Edible mushrooms: Role in the prevention of cardiovascular diseases". *Fitoterapia, 81*, 715-23.

Manchón N, D'Arrigo M, García-Lafuente A, Guillamón E, Villares A, Ramos A, Martínez J A & Rostagno M A (2010). "Fast analysis of isoflavones by high-performance liquid chromatography using a column packed with fused-core particles". *Talanta, 82*, 1986-94.

Rostagno M A, Villares A, Guillamon E, Garcia-Lafuente A & Martinez J A (2009). "Sample preparation for the analysis of isoflavones from soybeans and soy foods". *Journal of Chromatography A, 1216*, 2-29.

García-Lafuente A, Guillamón E, Villares A, Rostagno M A & Martínez J A (2009). "Flavonoids as anti-inflammatory agents: Implications in cancer and cardiovascular disease". *Inflammation Research, 58,* 537-52.

In: Biochemistry Researcher Biographical Sketches ... ISBN: 978-1-62081-308-9
Editors: J. T. Walker and C. M. Jones

Chapter 23

Vesna Vucic

Affiliation:

Centre of Research Excellence in Nutrition and Metabolism, Institute for Medical Research, University of Belgrade, Belgrade, Serbia

Address:

Tadeusa Koscuska 1, 11129 Belgrade, Serbia

Date of Birth: 25/07/1971

Education:

BSc Biochemistry, MSc Biochemistry, PhD Biology (field: Biochemistry and Physiology)

Research and Professional Experience:

1995-2000. Institute for Application of Nuclear Energy, Zemun, Serbia;
2001-2007 Vinca institute for nuclear sciences, Belgrade, Serbia;
2007-presentÂ Institute for Medical Research, Belgrade, Serbia

Professional Appointments:

1995-2000. Research assistant at Institute for Application of Nuclear Energy, Zemun, Serbia;
2001-2007 Research Associate at Vinca institute for nuclear sciences, Belgrade, Serbia;
2007-present Research Associate Professor at Institute for Medical Research, Belgrade, Serbia and 2008-present involved in FP6 project Eurreca (European Micronutrient Recommendations Aligned) Network of Excellence

Publications Last 3 Years:

1. Arsi, Vu, Tepši J, Mazi S, Djeli M, Glibeti M. Altered plasma and erythrocyte phospholipid fatty acid profile in elite female water polo and football players. *Appl Physiol Nutr Metab.* 2012 Feb;37(1):40-7. Epub 2011 Dec 13.
2. Popovic T., Borozan S., Arsic A., Martacic-Debeljak J., Vucic V., Trbovic A., Mandic LJ., Glibetic M. Fish oil supplementation improved liver phospholipids fatty acid composition and parameters of oxidative stress in male wistar rats. *Journal of Animal Physiology and Animal Nutrition*, DOI: 10.1111/j.1439-0396.2011.01216.x.
3. Tepsic J., Vucic V., Arsic A., Mazic S., Djelic M., Glibetic M. Unfavourable plasma and erythrocyte phospholipid fatty acid profile in elite amateur boxers. *European Journal of Sport Science*, 2011 in press.
4. Â Cvetkovic Z, Vucic V, Cvetkovic B, Petrovic M, Risti Medi D, Tepši J, Glibeti M Abnormal fatty acid distribution of the serum phospholipids of patients with non-Hodgkin lymphoma. *Annals of Hematology*, 2010, 89:775-782.
5. Popovic T, Borozan S, Arsic A, Debeljak-Martacic J, Vucic V, de Luka S, Milanovic I, Trbovic A, Glibetic M. Effects of n-3 Supplementation on Plasma and Liver Phospholipid Fatty Acids Profile in Aged Wistar Rats. *Croat Chem Acta*, 2011, 84:73-79.
6. Tepsic J., Vucic V., Arsic A., Blazencic-Mladenovic V., Mazic S., Glibetic M. Plasma and erythrocyte phospholipid fatty acid profile in professional basketball and football players. *European Journal of Applied Physiology*, 2009,107:359-365.

In: Biochemistry Researcher Biographical Sketches … ISBN: 978-1-62081-308-9
Editors: J. T. Walker and C. M. Jones

Chapter 24

Paulus S. Wang

Affiliation: Professor, Ph.D

Address:

Department of Physiology, School of Medicine, National Yang-Ming University, Shih-Pai, Taipei, Taiwan (11221), Republic of China.

Date of Birth: July 9, 1949

Education:

1972 - B.S. Department of Zoology, National Taiwan University, Taipei, Taiwan, R.O.C.
1974 - M.S. Department of Zoology, National Taiwan University. Taipei, Taiwan, R.O.C.
1980 - Ph.D. Department of Veterinary Biosciences, University of Illinois at Champaign-Urbana, Illinois, U.S.A.

Research and Professional Experience:

1974/8-1975/7 Assistant Research Fellow, Institute of Zoology, Academia Sinica, Taipei, Taiwan, R.O.C.
1980/4-1988/7 Associate Professor, Department of Physiology, National Yang-Ming Medical College, Taipei, Taiwan, R.O. C.
1985/9-1986/9 Fogarty International Fellow, Department of Obstetrics and Gynecology, The University of Texas Health Science Center at Dallas, Dallas, Texas, U.S.A.
1986/10-1992/7 Chairman, Department and Institute of Physiology, National Yang-Ming Medical College, Taipei, Taiwan, R.O.C.
1988/8-1994/7 Professor, Department and Institute of Physiology, National Yang-Ming Medical College, Taipei, Taiwan, R.O.C.

1994/8-Present Professor, Department and Institute of Physiology, National Yang-Ming University, Taipei, Taiwan, R.O.C.
2002/3-2005/1 Vice-Dean, School of Medicine, National Yang-Ming University, Taipei, Taiwan, R.O.C
2005/7-2009/12 Director, Laboratory Animal Center, National Yang-Ming University, Taipei, Taiwan, R.O.C.

Honors:

1985 International Research Fellowship (No.1F05TW03487-01) from the Fogarty International Center of the National Institutes of Health, U.S.A.
1987 Award by the Ministry of Education for Teaching Materials
1988 Travel Award, International Society of Endocrinology
1988 Research Chair Award from the Cjing-Ling Medical Foundation, R.O.C.
1988-1992 Excellent Research Award From the National Science Council, R.O.C.
1991 Parke-Davis Award of Distinguished Paper, The Endocrine Society of The Republic of China.
1996 Outstanding Research Award from the Chjing-Ling Medical Foundation, R.O.C.
2000-2001 Outstanding Research Award from the National Science Council, R.O.C.

Publications Last 3 Years:

1. Chang, L.L., Wun, W.S.A., and Wang, P.S. Mechanism of inhibition of dehydroepiandrosterone upon corticosterone release from rat zona fasciculata-reticularis cells. *J. Cell. Biochem.* 140:359-368, 2008.
2. Lee, S.D., Kou, W.W., Bau, D.T., Ko, F.Y., Wu, F.L., Kou, C.H., Hwang, J.M., Wang, P.S., Lu, M.C., and Huang, C.Y. The coexistence of nocturnal sustained hypoxia and obesity additively increase cardiac apoptosis. *J. Appl. Physiol.* 104:1144-1153, 2008.
3. Wu, C.L., Doong, M.L., and Wang, P.S. Involvement of cholecystokinin receptor in the inhibition of gastrointestinal motility by oxytocin in ovariectomized rats. *Eur. J. Pharmacol.* 580:407-415, 2008.
4. Chang, L. L., Wun, W. S. A., and Wang, P. S Effects of dehydroepiandrosterone on aldosterone release in rat zona glomerulosa cells. *J. Biomed. Sci.* 15:463-470, 2008.
5. Wang, S.W., Hwang, G.S., Chen, T.J., and Wang, P.S. Effects of arecoline on testosterone release in rats. *Am. J. Physiol.* (Endocrinol. Metab.) 295: E497-E504, 2008.
6. Hisa, S.M., Kuo, Y.H., Chiang, W., and Wang, P.S. Effects of adlay (*Coix lachryma-jobi* L. var. *ma-yuen* Stapf.) hull extracts on uterine contraction and Ca^{2+} mobilization in the rat. *Am. J. Physiol.* (Endocrinol. Metab.) 295: E719-E726, 2008.
7. Yu, C.H., Kan, S.F., Pu, H.F., Jea-Chien, E., and Wang, P.S. Apoptic signaling in bufalin- and cinobufagin-treated androgen-dependent and –independent human prostate cancer cells. *Cancer Sci.* 99: 2467-2476, 2008.
8. Kau, M.M., Kau, S.F., Wang, J.R., Wang, P.S., Lau, Y.T. and Wang, S.W. Acute effects of digoxin on plasma aldosterone and cortisol in monkeys. *Metabolism* 58: 55-61, 2009.

9. Yu, C.H., Kan, S.F., Shu, C.H., Lu, T.J., Sun-Hwang, L., and Wang, P.S. Inhibitory mechanisms of *Agaricus blazei Murill* on the growth of prostate cancer in vitro and in vivo. *J. Nutr. Biochem.* 20: 753-764, 2009.
10. Hsia, S.M., Tseng, Y.W., Wang, S.W., Kuo, Y.H., Huang, D.W., Wang, P.S.* and Chiang, W. Effects of adlay (*Coix lachrymal-jobi L. var, ma-yuen Stapf.*) hull extracts on testosterone release from rat Leydig cells. *Phytother. Res.* 23: 687-695, 2009.
11. Feng, M., Qin, J., Wang, C., Ye, Y., Wang, S., Xie, D., Wang, P.S. and Liu, C. Estradiol upregulates the expression of oxytocin receptor in colon in rats. Am. J. Physiol. *Gdocrinol. Metab.* 296: E1059-E1066, 2009.
12. Qin, J., Feng, M., Wang, C., Ye, Y., Wang, P.S. and Liu, C.* Oxytocin receptor expressed on the smooth muscle mediates the excitatory effect of oxytocin on gastric motility in rats. *Neurogastroenterol. Motil.* 21: 430-438, 2009.
13. Lu, C.L., Hsieh, J.C., Dun, N.J., Oprea, T.I., Wang, P.S., Luo, J.C., Lin, H.C., Chang, F.Y. and Lee, S.D. Estrogen rapidly modulates 5-hydroxy tryptophan-induced visceral hypersensitivity in awake female rat through GPR30 receptors. *Gastroenterology* 137: 1040-1050, 2009.
14. Yu, P.L., Chao, H.L., Wang, S.W., and Wang, P.S. Effects of evodiamine and rutaecarpine on the secretion of corticosterone by zona fasciculata-reticularis cells in male rats. *J. Cell. Biochem.* 108: 469-475, 2009.
15. Yu, P.L., Wu, C.I., Lee, T.S., Pan, W.H.T., Wang, P.S. and Wang, S.W. Attenuation of estradiol on the reduction of striatal dopamine by amphetamine in ovariectomized rats. *J. Cell. Biochem.* 108: 1318-1324, 2009.
16. Huang, W.J., Hung, C.R., Chen, M.C.,Wang, J.R., Doong, M.L., Wang, C.K., Hung, C.T., and Wang, P.S. Involventment of cyclooxygenase 2 and prostaglandin E_2 in the effects of insulin on gastric emptying in male rats. *J. Physiol. Pharmacol.* 60: 109-118, 2009.
17. Liu, T.C., Kuo, C.H. and Wang, P.S.* Exercise and testosterone. *Adapt. Med.* 1:26-31, 2009.
18. Chang, H.C., Yang, Y.R., Wang, P.S. and Wang, R.Y. Effects of treadmill training on motor performance and extracellular glutamate level in striatum in rats with or without transient middle cerebral artery occlusion. *Behav. Brain Res. 205*: 450-455, 2009.
19. Hsia, S. M., Su, J. C., Wang, K. L., and Wang, P.S. Effects of evodiamine on the expression and secretion of vascular endothelial growth factor in human breast cancer cells. *Clin. Mol. Med.* 1: 8-13, 2009.
20. Wu, J.J., Wang, K.L., Hwang, G.S., Mao, I.F., Chen, M.L., and Wang, P.S. Differential effects of nonylphenol on testosterone secretion in rat Leydig cells. *Toxicology* 268: 1-7, 2010.
21. Yu, P.L., Pu, H.F., Chen, S.Y., Wang, S.W., and Wang, P.S. Effects of catechin, epicatechin and epigallocatechin gallate on testosterone production in rat Leydig cells. *J. Cell. Biochem.* 110: 333-343, 2010.
22. Wu, J.J., Wang, K.L., Mao, I.F., Chen, M.L., and Wang, P.S. Effects of oral nonylphenol on testosterone production in rats Leydig cells. *Adapt. Med.* 2: 47-52, 2010.
23. Tsague-Manfo, F.P., Chao, W.F., Fewou-Moundipa, P., Pugeat, M., and Wang, P.S. Effects of maneb on testosterone release in male rats. *Drug Chem. Toxicol.* 00:000-000, 2010. (In Press)

24. Chang, H.C., Yang, Y.R., Wang, P.S., Kuo, C.H., and Wang, R.Y.* Effects of insulin-like growth factor 1 on muscle atrophy and motor function in rats with brain ischemia. *Chinese J. Physiol*. 00: 000-000, 2010. (In Press)
25. Chen, M.C., Lin, H., Hsu, F.N., Huang, P.H., Lee, G.S., and Wang, P.S. Involvement of cAMP in nerve growth factor-triggered p35/Cdk5 activation and differentiation in PC12 cells. *Am. J. Physiol. Cell Physiol*. 299: 000-000, 2010. (In Press)
26. Chen, M.C., Yu, C.H., Wang, S.W., Pu, H.F., Kan, S.F., Lin, L.C., Chi, C.W., Ho, L.L.T., Lee, C.H., and Wang, P.S. Anti-proliferative effects of evodiamine on human thyroid cancer cell line ARO. *J. Cell. Biochem*. 110: 1495-1503, 2010.
27. Hsia, S. M. and Wang, P. S.* Effects of adlay extract on steroidogenesis and uterine contraction. *Adapt. Med*. 2: 104-107, 2010.
28. Chang, Y.J., Chen, M. C., Chang, F. Y., and Wang, P. S. Radioimmunoassay for ghrelin: evaluation of method and the effect of fasting in rats. *Adapt. Med*. 2: 108-112, 2010.
29. Cho, Y.M., Pu, H.F., Huang, W.J., Ho, L.T., Wang, S.W., and Wang, P.S. Role of serum- and glucocorticoid- inducible kinase-1 in regulating torsion-induced apoptosis in rats. *Int. J. Androl*. xxx: 000-000, 2010. (In Press)
30. Hsu, R.L., Jing, C.J., and Wang, P.S. Chronic effects of propylthiouracil on the metabolism and secretion of adrenocorticotropin and corticosterone in male rats. *J. Med. Sci*. 00: 000-000, 2010. (In Press)
31. Huang, W.J.S., Yang L.Y., Pu, H.F., Tsai, Y.T. and Wang. P.S. The effects of anti-TNF-α antibody on hyperprolactinemia-related suppression of hCG-induced testosterone release in male rats. *J. Sex. Med*. xx: xxx-xxx, 2010. (In Press)
32. Chang, L.L., Wun, W.S.A., and Wang, P.S.* Effects and mechanisms of nonylphenol on corticosterone release in rat zona fasciculata-reticularis cells. *Toxicol. Sci*. xxx xxx-xxx, 2010. (In Press)
33. Chen, M.C., Wang, S.W., Kan, S.F., Tsai, S.C., Wu, Y.C., and Wang, P.S. Stimulatory effects of propylthiouracil on pregnenolone production through upregulation of steroidogenic acute regulatory protein expression in rat granulosa cells. *Toxicol. Sci*. xxx:xxx-xxx, 2010. (In Press)
34. Hsu, Y.W., Pan, Y.J., Cho, Y.M., Liou, T.H., Chou, P. and Wang, P.S. Aging effects on exercise-induced alternations in plasma acylated ghrelin and leptin in male rats. Eur. *J. Appl. Physiol*. 000: xxx-xxx, 2010. (In Press)
35. Yang, Y.R., Chang, H.C., Wang, P.S. and Wang, R.Y. Motor performance improved by exercise in cerebral ischemic rats. *Scand. J. Med. Sci. Sports* (Submitted)

In: Biochemistry Researcher Biographical Sketches … ISBN: 978-1-62081-308-9
Editors: J. T. Walker and C. M. Jones

Chapter 25

William R. Ware

Affiliation:
Faculty of Science (Emeritus), University of Western Ontario, London, Canada

Address:
14 Metamora Cres. London, Ontario, Canada, N6G 1R3

Date of Birth: June 13, 1931

Education:

Ph.D., University of Rochester, Nuclear and Physical Chemistry

Research and Professional Experience:

Industrial and academic in the fields of photochemistry, photophysics and photobiology

Professional Appointments:

Professor of Chemistry, University of Minnesota, then Professor of Chemistry (now Emeritus), University of Western Ontario

Honors:
Fellow of the Chemical Society of Canada

Publications Last 3 Years:

1. Ware WR. The JUPITER lipid lowering trial and vitamin D: Is there connection? Dermatoendocrinol. 2010 Apr;2(2):50-4. PubMed PMID:21547097;
2. Ware WR. The mainstream hypothesis that LDL cholesterol drives atherosclerosis may have been falsified by non-invasive imaging of coronary artery plaque burden and progression. Med Hypotheses. 2009 Oct;73(4):596-600. Epub 2009. Jun 26. PubMed PMID: 19560285.
3. Ware WR. Nutrition and the prevention and treatment of cancer: association of cytochrome P450 CYP1B1 with the role of fruit and fruit extracts.Integr Cancer. Ther. 2009 Mar;8(1):22-8. Epub 2008 Dec 30. Review. PubMed PMID: 19116217.
4. Ware WR. P450 CYP1B1 mediated fluorescent tumor markers: a potentially useful approach for photodynamic therapy, diagnosis and establishing surgical margins. Med Hypotheses. 2009 Jan;72(1):67-70. Epub 2008 Oct 1. PubMed PMID:18829176.
5. Ware WR. Psychological stress, insulin resistance, inflammation and the assessment of heart disease risk. Time for a paradigm shift? Med Hypotheses. 2008;71(1):45-52. Epub 2008 Apr 10. PubMed PMID: 18406066.
6. Ware WR. High cholesterol and coronary heart disease in younger men: the potential role of stress induced exaggerated blood pressure response. Med Hypotheses. 2008;70(3):543-7. Epub 2007 Aug 21. PubMed PMID:17714881.

Invited Chapter:

Ware,W., 2009. Is Coronary Heart Disease Risk Underestimated in the Primary Care Setting? Potential Importance of Psychological Stress Assessment. In: L.Sher (Editor), Pshchological Factors and Cardiovscular Disorders. Nova Biomedical Books, New York, pp. 351-380.

In: Biochemistry Researcher Biographical Sketches ... ISBN: 978-1-62081-308-9
Editors: J. T. Walker and C. M. Jones

Chapter 26

Anna Watson

Affiliation: Baker IDI Heart and Diabetes Institute

Address:

c/o Baker IDI Heart and Diabetes Institute, PO Box 6492, St.Kilda Road Central, Melbourne VIC 8008, Australia

Date of Birth: 10/11/1975

Education: BSc(Hons), PhD

Research and Professional Experience:

My research into the neural and vascular effects of peptide hormones includes experience in 5 different laboratories in 3 different countries with this research experience resulting in peer-reviewed publications from projects in 4 of these laboratories (22 publications to date IFs 8-3).

In 2006 I commenced my postdoctoral position in the Diabetes Complications Division at the Baker IDI Institute where I have supervised and/or mentored 9 students and researched the role peptide hormones such as endothelin I, angiotensin II, urotensin II, and also advanced glycation end products in the development of diabetes complications, specifically nephropathy and atherosclerosis. Recently I have been awarded a Baker IDI Early Career grant with toward my ongoing research into the vascular actions of the peptide urotensin II in diabetes.

Professional Appointments:

- National Health and Medical Research Council of Australia Postdoctoral Biomedical Fellow (2006-current), Baker IDI Heart and Diabetes Institute, Melbourne, VIC Australia
- Research Officer (2005-2006) Howard Florey Institute, The University of Melbourne, VIC Australia
- PhD (2001-2005), Howard Florey Institute, The University of Melbourne, VIC Australia
- Guest Scientist (2000) German Institute of Human Nutrition (DIfE), Potsdam, Germany
- Bio-Medical Technician (2000) Institute of Ophthalmology, Department of Pathology, London, UK
- Technical Officer (1998-2000) Howard Florey Institute, The University of Melbourne, VIC Australia
- Honors: (1997) Department of Zoology, The University of Melbourne, VIC Australia

Publications Last 3 Years:

1. Watson, A.M.D., Sheehy K, Li J, Calkin A. C., Koitka A., Rajan, S.N., Brasacchio D., Allen, T.J., Cooper M. E., Thomas M. C., Jandeleit-Dahm K. J. A (2011) Delayed intervention with AGE inhibitors attenuates the progression of diabetes-accelerated atherosclerosis in diabetic apolipoprotein E knockout mice. *Diabetelogia* 54(3):681-689.
2. Wang, B., Koh, P., Winbanks, C., Coughlan, M. T., McClelland, A., Watson, A.M.D., Jandeleit-Dahm, K., Burns, W. C., Thomas, M. C., Cooper, M. E. and Kantharidis, P. (2011), miR-200a Prevents renal fibrogenesis through repression of TGF-beta2 expression, *Diabetes*, 60: 280-287.
3. Watson, A., Thomas, M., Koh, P., Figorala, J., Rahbar, S. and Jandeleit-Dahm, K., Attenuation of diabetes associated atherosclerosis with LR-90, a novel inhibitor of AGE formation, In: Thomas, M. and Forbes, J. (eds), *Maillard Reaction. Interface between Aging, Nutrition and Metabolism,* Royal Society of Chemistry Publishing, London, UK, 2010: 137-143.
4. Watson, A.M.D., Soro-Paavonen, A., and Jandeleit-Dahm K.A.M. AGE-RAGE Signalling and Endothelial Dysfunction and Atherosclerosis in Diabetes. In: Duphinee SM, Karsan A (eds) *Endothelial Dysfunction and Inflammation*. Birkh¤user-Verlag/Springer, Basel Switzerland, 2010: 161-174.
5. Wang B, Herman-Edelstein M, Koh P, Burns W, Jandeleit-Dahm K, Watson A, Saleem M, Goodall GJ, Twigg SM, Cooper ME, Kantharidis P (2010). E-cadherin expression is regulated by miR-192/215 by a mechanism that is independent of the profibrotic effects of TGF{beta}. *Diabetes* 59(7):1794-1802.Ramchandra, R., Watson, A.M.D., Hood, S.G. & May, C.N. (2010) The response of cardiac sympathetic nerve activity to intravenous irbesartan in heart failure. *Am J Physiol Regul Integr Comp Physiol* 298(4):R1056-1060.

6. Watson, A.M.D., Jiaze, L., Schumacher, C., de Gasparo, M., Feng, B., Thomas, M.C., Allen, T.J., Cooper, M.E., Jandeleit-Dahm, K.A.M. (2010) The endothelin receptor antagonist avosentan ameliorates nephropathy and atherosclerosis in diabetic apolipoprotein E knockout mice. *Diabetologia* 53:192-203.
7. Pham, Y., Tu, Y., Wu, T., Allen, T.J., Calkin, A.C., Watson, A.M.D, Li, J., Jandeleit-Dahm K.A., Toh, B-H, Cao, Z., Cooper, M.E., Chai, Z. (2010) Cell division autoantigen 1 plays a profibrotic role by modulating downstream signalling of TGF-Î² in a murine diabetic model of atherosclerosis. *Diabetologia* 53:170-179KoÃ¯tka, A., Cao, Z., Koh, P., Watson, A.M.D., Sourris K.C., Loufrani, L., Soro-Paavonen A., Walther T., Woollard K.J., Jandeleit-Dahm, K.A.M, Cooper, M.E., and Allen T.J. (2010) Angiotensin type 2 receptor blockade and deficiency attenuate the development of atherosclerosis in an apolipoprotein E-deficient mouse model of diabetes. *Diabetologia* 53(3):584-592.
8. Watson, A.M.D., McKinley, M.J., May, C.N. (2008) Effect of central urotensin II on heart rate, blood pressure and brain Fos immunoreactivity in conscious rats. *Neuroscience* 155:241-249. IF: 3.556Ramchandra R, Hood S.G., Watson A.M.D., May C.N. (2008) Responses of cardiac sympathetic nerve activity to changes in circulating volume differ in normal and heart failure sheep. *Am J Physiol Regul Integr Comp Physiol.* 295:R719-726. IF: 2.196.
9. Soro-Paavonen A., Watson A.M.D., Li J., Paavonen K., Koitka A., Calkin A.C., Barit D., Coughlan M.T., Drew B.G., Lancaster G.I., Thomas M., Forbes J.M., Nawroth P.P., Bierhaus A., Cooper M.E., Jandeleit-Dahm K.A.M. (2008) RAGE deficiency attenuates the development of atherosclerosis in diabetes. *Diabetes* 57:2461-2469. IF: 8.398.
10. Jandeleit-Dahm, K., Watson, A., Soro-Paavonen, A. (2008) The AGE/RAGE axis in diabetes-accelerated atherosclerosis. *Clin Exp Pharmacol Physiol.* 35:329-334. IF: 2.196.

In: Biochemistry Researcher Biographical Sketches ... ISBN: 978-1-62081-308-9
Editors: J. T. Walker and C. M. Jones

Chapter 27

Lei Xu

Affiliation: University of Minnesota

Address:
203 Kaufert Lab, 2004 Folwell Ave., St. Paul, MN 55108

Date of Birth: March 26th

Education: PhD

Research and Professional Experience:

3 years of encapsulation experience
Over 1 year of industrial process engineer experience
5 years of environmental microbiology culture experience
3 years in biomass pretreatment research
More than 2 years of gene cloning, sequencing, expression and purification experience
2 years in bio-catalysis and biofuel fermentation process optimization.

Professional Appointments: Instructor

Publications Last 3 Years:

Lei Xu and Ulrike Tschirner, Improved ethanol production through anaerobic thermophilic co-culture from various carbohydrates. Extended abstract in Proceeding of the AICHE's 2011 Annual Meeting.

Lei Xu and Ulrike Tschirner, Preacetic acid pretreatment of alfalfa stem and aspen biomass. *Bioresource* (Accepted for publication in 2011).

Lei Xu and Ulrike Tschirner, Improved ethanol production from various carbohydrates through anaerobic thermophilic co-culture. *Bioresource Technology*. 2011; 102 (21):10065-10071.

Lei Xu, Ulrike Tschirner and Waleed Al-dajani. Comparing impact of pretreatment of aspen and alfalfa biomass on fermentable sugar yield. August, 2008. International Bioenergy and Bioproducts Conference. Portland, Oregon, USA.

Lei Xu and Ulrike Tschirner, Optimizing pretreatment conditions for biomass conversion. Sixth Annual Lifescience Alleys Conference and Expo, St. Paul, MN, December, 2007.

Lei Xu and Ulrike Tschirner, Improvement of ethanol production from various carbohydrates by an anaerobic thermophilic co-culture. March, 2011. International Bioenergy and Bioproducts Conference. Atlanta, Georgia, USA.

Lei Xu and Ulrike Tschirner. Comparing Impact of Pretreatment of Aspen and Alfalfa Biomass on Fermentable Sugar Yield, 2008 Int'l. Bioenergy & Bioproducts Conference. Portland, Oregon, August, 2008.

In: Biochemistry Researcher Biographical Sketches ... ISBN: 978-1-62081-308-9
Editors: J. T. Walker and C. M. Jones

Chapter 28

Oguzhan Yildiz

Affiliation: Assoc. Professor of Pharmacology

Address:

Department of Medical Pharmacology, Gülhane School of Medicine, 06018, Ankara, Turkey Telephone: +90 312 3044766; FAX:+90 312 3042700

Date of Birth: 4 May 1963

Education:

Gulhane School of Medicine (medical education),
Hacettepe University Faculty of Medicine Ankara (Doctorate program),
University of California Irvine, CA, USA (one year of sabbatical visit, 1997)

Research and Professional Experience:

16 years of lecturing on pharmacology at the Department of Pharmacology, Gülhane School of Medicine. His recent research interests include non-genomic vascular effects of testosterone, oxidative stress and endothelium. He has more than 50 scientific papers in journals indexed in SCI.

Professional Appointments:

Between years 1991-1993, he worked in Department of Pharmacology, Hacettepe University Faculty of Medicine, Ankara.

In 1996, he became an Associate Professor in Department of Medical Pharmacology, Faculty of Medicine, Gulhane Military Medical Academy, Ankara.

In 1997, he worked in Department of Pharmacology, Faculty of Medicine, University of California Irvine, USA.

Honors:

1994 Young Investigator Award, 3rd International Satellite Meeting on Serotonin, Chicago, USA
1995 Young Investigator Award, EPHAR-95, Milan, Italy
1995 First Prize, The Best Research Award, National Congress of Turkish Optalmological Society
1996 First Prize, The Best Poster Presentation Award, 13rd National Congress of Pharmacology
1996 First Prize, Research Awards, Turkish Diabetes Association Prof. Celal Öker Foundation
1996 First Prize, Novartis Scentific Research Awards in Pharmacology
1997 Third Prize, Research Awards, Turkish Diabetes Association Prof. Celal Öker Foundation
1998 First Prize, National Medical Research Awards, 100th Anniversary of the Foundation of Gülhane Military Medical Academy
1999 Best Scientific Publication Prize, Turkish Diabetes Association Prof. Celal Öker Foundation
2000 Best Scientific Research Project Prize, Turkish Diabetes Association Prof. Celal Öker Foundation
2000 Second Prize, Novartis Scentific Research Awards in Pharmacology
2001 Best Scientific Publication Prize, Turkish Diabetes Association Prof. Celal Öker Foundation
2001 Best Scientific Publication Prize, Turkish Diabetes Association Prof. Celal Öker Foundation
2002 Third Prize, Gulhane School of Medicine, Internal Medicine Scentific Study Awards
2003 First Prize, Presenations, Cukurova Colo-proctology Symposium
2004 Second Prize, Novartis Scentific Research Awards in Pharmacology
2005 Second Prize, Gulhane School of Medicine, Internal Medicine Scentific Study Awards
2008 First Prize, Oral Presentations, 20. National Urology Congress
2008 First Prize, Gulhane School of Medicine, Institute of Medical Sciences Scentific Study Awards
2008 First Prize, Gulhane School of Medicine, Institute of Medical Sciences Scentific Study Awards
2009 Second Prize, Gulhane School of Medicine, Surgical Sciences Scentific Study Awards
2010 Second Prize, Oral Presentations, 12. National Internal Medicine Congress

Publications Last 3 Years:

1. Yildiz O, Seyrek M. Vasodilating mechanisms of testosterone. *Exp Clin Endocrinol Diabetes*. 2007 Jan;115(1):1-6. Review.
2. Seyrek M, Yildiz O, Ulusoy HB, Yildirim V. Testosterone relaxes isolated human radial artery by potassium channel opening action. *J Pharmacol Sci*. 2007 Mar;103(3):309-16.
3. Yildiz O. Vascular Smooth Muscle and Endothelial Functions in Aging. *Ann N Y Acad Sci*. 2007 Apr;1100:353-60. Review
4. Tunca YM, Aydın C, Ozen T, Seyrek M, Ulusoy HB, Yildiz O. Effect of Mineral Trioxide Aggregate (MTA) on the Contractility of the Rat Aorta. *J Endodon*. 2007 Jul;33(7):823-6.
5. Yildiz O. Vasodilatory Mechanisms of Levosimendan; Involvement of K+ Channels. *J Pharmacol Sci*. 2007 May;104(1):1-5. Review.
6. Uygun A, Ozel AM, Yildiz O, Aslan M, Yesilova Z, Erdil A, Bagci S, Gunhan O.Comparison of three different second-line quadruple therapies including bismuth subcitrate in Turkish patients with non-ulcer dyspepsia who failed to eradicate Helicobacter pylori with a 14-day standard first-line therapy. *J Gastroenterol Hepatol*. 2007.
7. Yildiz O, Ulusoy HB, Seyrek M, Gul H, Yildirim V.Dexmedetomidine produces dual alpha2-adrenergic agonist and alpha1-adrenergic antagonist actions on human isolated internal mammary artery. *J Cardiothorac Vasc Anesth*. 2007 Oct;21(5):696-700.
8. Odabası E, Gul H, Macit E, Turan M, Yildiz O. Lipophilic components of different therapeutic mud species. *J Altern Complement Med*. 2007 Dec;13(10):1115-8.
9. Ulusoy HB, Gul H, Seyrek M, Yildiz O, Ulku C, Yıldırım V, Kuralay E, Celik T, Yanarates O. The Concentration- Dependent Contractile Effect of Methylene Blue in Human Internal Mammary Artery: A Quantitative Approach to Its Use in Vasoplegic Syndrome. *J Cardiothorac Vasc Anesth*. 2008 Aug;22(4):560-4.
10. Irkilata HC, Yildiz O, Yildirim I, Seyrek M, Basal S, Ulku C, Dayanc M. The vasodilator effect of testosterone on human internal spermatic vein and its relation with the grade of varicocele. *J. Urology* 2008 Aug;180(2):772-6.
11. Basak F, Vural IM, Kaya E, Ulku C, Guven G, Cehreli SB, Cehreli ZC, Seyrek M, Yildiz O. Vasorelaxant effect of a self-etch adhesive system through calcium antagonistic action. *J Endod*. 2008 Oct;34(10):1202-6.
12. Satar B, Yildiz O, Karatas E, Boydag S, Akkaya A. Pentoxifylline response in alloxan-induced diabetic rats. *Acta Oto-laryngologica* 2008 Nov;128(11):1221-6.
13. Yildiz O, Seyrek M, Irkilata HC, Yildirim I, Tahmaz L, Dayanc M.Testosterone might cause relaxation of human corpus cavernosum by potassium channel opening action. *Urology*. 2009 Jul;74(1):229-32.
14. Seyrek M, Guven G, Vural IM, Kaya E, Cehreli ZC, Altun C, Yildiz O. Vasorelaxant Effects of Clearfil SE Bond and Clearfil S3 Bond Mediated by Calcium Antagonistic Action. *Journal of Biomaterials Science*. 2009 (baskıda).
15. Celik S, Oktenli C, Terekeci HM, Ipcioglu O, Sanisoglu YS, Sayan S, Yesilova Z, Yildiz O, Tunca Y, Nalbant S. Blood oxidative stress biomarkers in patients with familial mediterranean fever. *Akt Rheumatol.* 2009; 34:1-4.

16. Ozen T, Kilic A, Bedir O, Koru O, Sorkun K, Tanyüksel M, Kilic S, Gencay O, Yildiz O, Baysallar M. In Vitro Activity of Turkish Propolis Samples Against Anaerobic Bacteria Causing Oral Cavity Infections. *Kafkas Univ Vet Fak Derg* 2010: 16(2): 293-298.
17. Seyrek M, Vural IM, Tunca YM, Aydin C, Ulku C, Demirkaya K, Inal A, Yildiz O. The Vasodilatory Effect of a Synthetic Polymer-based Root Canal Material on Rat Thoracic Aorta. *Int Endod J.* 2010: 43(7):590-9.
18. Karacayli U, Vural IM, Seyrek M, Yildiz O. Vasocontractile effect of SeraSeal in rat thoracic aorta. 2010: 31(9):1064-5.
19. Dogrul A, Gul H, Yesilyurt O, Ulas UH, Yildiz O. Systemic and spinal administration of etanercept, a tumor necrosis factor alpha inhibitor, blocks tactile allodynia in diabetic mice. *Acta Diabetol.* 2010 Nov 23. [Epub ahead of print]

Part 2 - Research Summaries on Biochemistry

In: Biochemistry Researcher Biographical Sketches ... ISBN: 978-1-62081-308-9
Editors: J. T. Walker and C. M. Jones

Chapter 29

57Fe Mössbauer Spectroscopic and Density Functional Theory (DFT) Study on the Interactions of the Metal Ion with Monosaccharides

Yassin Jeilani, Beatriz H. Cardelino and Natarajan Ravi
Spelman College, Atlanta, GA, US

Research Summary

The fundamental electrical and magnetic interactions between iron ions and biomaterials were investigated using an experimental approach, 57Fe Mössbauer spectroscopy, and ab initio computational methods. A conventional spin-Hamiltonian approach adopted for the data analysis of the Mössbauer data showed that the metal ion in the Fe-chitosan complex is in the high-spin ferric state and that it has an internal magnetic field of approximately 440 kG, at the nucleus. The magnitude of the internal field arises from the predominant Fermi-contact interaction of the high-spin ferric species with N/O ligands. This book proposes a scheme for the Fe-chitosan complex based on the analysis of the experimental data.

In: Biochemistry Researcher Biographical Sketches ...
Editors: J. T. Walker and C. M. Jones
ISBN: 978-1-62081-308-9

Chapter 30

An Introductory Course on Group Theory and Chemical Applications

Hikmat S. Hilal and Abed Al-Hafez Sayda
An-Najah N. University, West Bank, Palestine

Research Summary

Group theory is conventionally taught in a one-semester course at the senior/graduate level for science students. For such a purpose, many textbooks, dealing with rigors of derivations and theorems, have been written to give in-depth treatment. For one reason or another, scientists with no prior knowledge of group theory may need quick familiarity with the subject, without mathematical derivation rigors.

This book has been written for this purpose. The book is based on lecture notes given to senior/graduate students who need to deal with bonding theories and spectroscopy, without pre-knowledge of group theory.

In: Biochemistry Researcher Biographical Sketches ... ISBN: 978-1-62081-308-9
Editors: J. T. Walker and C. M. Jones

Chapter 31

Aromatic Polyethers Based on Heterocyclic Monomers

Alexander L. Rusanov and Nataliya M. Belomoina
Russian Academy of Sciences, Moscow, Russia

Research Summary

Poly(arylene ethers) (PAEs) represent an important class of condensation polymers, which were used as a basis for obtaining a number of commercial thermoplasts possessing high performance characteristics. The main method used for the synthesis of PAEs is based on the polycondensation of alkali-metal bisphenolates with activated electrophilic monomers such as arylene dinitro compounds and arylene dihalide compounds. These processes belong to reactions of the nucleophilic aromatic polysubstitution type (SNAr2), involving etherification as the polymer forming process. This book presents and reviews research on aromatic polyethers.

In: Biochemistry Researcher Biographical Sketches … ISBN: 978-1-62081-308-9
Editors: J. T. Walker and C. M. Jones

Chapter 32

Bacteriophages and Biofilms: Ecology, Phage Therapy, Plaques

Stephen T. Abedon
The Ohio State University, Mansfield, OH, US

Research Summary

Bacteriophages (phages) are the viruses of bacteria and biofilms that represent a frequent niche for bacteria, where they are embedded in extensive extracellular polymeric substances (EPS) and can be structured into complex microcolonies. As a consequence of the resulting spatial structure and heterogeneity, phage-bacterial interactions within biofilms can be more complicated than those between phages and planktonic bacteria. This book presents and discusses research which provides a better understanding of the biology of phages interacting with biofilms.

In: Biochemistry Researcher Biographical Sketches ...
Editors: J. T. Walker and C. M. Jones
ISBN: 978-1-62081-308-9

Chapter 33

Beta-Lactamases

Jean-Marie Frère
Universite de Liege, Centre for Protein Engineering, Belgium

Research Summary

The activity of a bacterial enzyme ¡°able to degrade penicillin¡± had first been described in 1940, even before the exact structure of penicillin was elucidated and, by 1970, several enzymes had been purified to homogeneity, the amino acid sequence of a staphylococcal penicillinase was also known and that of its <i>Bacillus licheniformis</i> counterpart was well under way. By contrast, their catalytic mechanism remained quite mysterious. A Zn++ metallo-beta-lactamase (¦Â-lactamase II, BcII or ¦ÂII) had also been described as soon as 1967 and later purified. It was thus surprising that the first mechanistic information demonstrating the presence of a penicillin-binding serine residue was obtained with a penicillin-sensitive DD-peptidase rather than a ¦Â-lactamase. This seemed to open the floodgates and several class A ¦Â-lactamases were then rapidly shown to be active-site serine enzymes. This book presents current research in the study of beta-lactamases.

In: Biochemistry Researcher Biographical Sketches … ISBN: 978-1-62081-308-9
Editors: J. T. Walker and C. M. Jones

Chapter 34

Biflavonoids: Occurrence, Structural Features and Bioactivity

Andrew G. Mercader and Alicia B. Pomilio
Universidad de Buenos Aires, Argentina

Research Summary

Biflavonoids comprise a group of the flavonoid family that possess a variety of structures and biological activities of high relevance, such as anticancer, antibacterial, antifungal, antiviral, anti-inflammatory, antinociceptive, antioxidant, vasodilator and anticlotting. The chemistry of biflavonoids is very important in many fields of research, especially because these compounds are structurally different bioactive molecules with potential for biomedical applications. This book highlights the structural biflavonoid variability, rearrangements and different stereochemistry through about 470 structures distributed in some species of Angiosperms, Gymnosperms, ferns and mosses.

In: Biochemistry Researcher Biographical Sketches ...
Editors: J. T. Walker and C. M. Jones
ISBN: 978-1-62081-308-9

Chapter 35

Binary Aqueous and CO_2 Containing Mixtures and the Krichevskii Parameter

***Aziz I. Abdulagatov*[1], *Ilmutdin M. Abdulagatov*[1] *and Gennadii V. Stepanov*[2]**
[1] Thermophysical Properties Division,
National Institute of Standards and Technology, Boulder, Colorado
[2] Institute for Geothermal Problems of the Dagestan Scientific
Center of the Russian

Research Summary

All available sources of the critical curves data of binary aqueous solutions and binary carbon dioxide containing mixtures have been collected, evaluated, and compared with various data sources and with prediction and correlation methods in this book. Various techniques of the critical parameter measurements for pure fluids and fluid mixtures and their accuracy were reviewed. Results of the critical analysis of the uncertainty of each data source along with comparisons with other sources were presented. The reliability, quality, and consistency of each critical property data set were examined. The most reliable critical parameters for binary H2O+solute and CO2+solute systems were selected and recommended for future scientific and industrial applications. Moreover, special interest has been focused on the description of thermodynamic and structural properties of the dilute mixtures near the critical point of pure solvent (water and carbon dioxide) using the critical properties data (initial slopes of the critical curves, the shape of the critical lines).

In: Biochemistry Researcher Biographical Sketches … ISBN: 978-1-62081-308-9
Editors: J. T. Walker and C. M. Jones

Chapter 36

Bioactive Oligosaccharides: Production, Biological Functions and Potential Commercial Applications

Aneli M. Barbosa, Robert F. H. Dekker and Ellen C. Giese
Biorefining Research Institute, Lakehead University, Thunder Bay, ON, Canada

Research Summary

This book presents an overview on different pathways leading to the production of bioactive oligosaccharides for biotechnological applications. Mostly, these carbohydrate oligomers constitute a nutritional source of “fiber” (prebiotic) that is beneficial to bacterial growth in the lower distal part of the human intestinal tract promoting health, and general well-being. Oligosaccharides, like some of their polysaccharide counterparts, can induce innate immune responses, and this unique property has led to potential applications for their commercialization as immunoceuticals.

In: Biochemistry Researcher Biographical Sketches ... ISBN: 978-1-62081-308-9
Editors: J. T. Walker and C. M. Jones

Chapter 37

Biochemistry and Function of Antifreeze Proteins

Steffen P. Graether
University of Guelph, Guelph, Ontario, Canada

Research Summary

Antifreeze proteins, also known as thermal hysteresis proteins, ice binding proteins and ice structuring proteins, prevent the growth of ice crystals in several cold blooded organisms. First discovered in fish, they have also been found in insects, plants, fungi and bacteria. Antifreeze proteins cause the non-colligative depression of the freezing point of water, a property which has been exploited in the practical applications of antifreeze proteins such as improving the texture of ice cream, and could be used to extend the crop growing season or allow fish to thrive in cold waters. This book provides clear information on what is known about antifreeze proteins today and how to study them.

In: Biochemistry Researcher Biographical Sketches … ISBN: 978-1-62081-308-9
Editors: J. T. Walker and C. M. Jones

Chapter 38

Biochemistry of Prenatal Alcohol Exposure

***Ernst van Faassen*[1] *and Onni Niemelä*[2]**
[1] Utrecht University, Utrecht, Netherlands
[2] University of Tampere, Finland

Research Summary

This book reviews the biochemical pathways of ethanol and its metabolites in humans in the context of pregnancy. Special emphasis is devoted to the implications for fetal and neonatal development. The book also covers primary direct effects of ethanol as well as downstream effector pathways involving modulation of receptor and hormonal signaling, and the enzymatic status of the offspring. Special attention is given to the implications of maternal alcohol consumption for the nutritional status of compounds with recognized relevance for pregnancies, such as folate, unsaturated fatty acids and zinc.

In: Biochemistry Researcher Biographical Sketches ... ISBN: 978-1-62081-308-9
Editors: J. T. Walker and C. M. Jones

Chapter 39

Biodegradation of Cellulose Fibers

***Barbara Simonèiè*[1], *Brigita Tomšiè*[1], *Boris Orel*[2] *and Ivan Jerman*[2]**

[1] University of Ljubljana, Faculty of Natural Sciences and Engineering, Department of Textiles, Ljubljana, Slovenia

[2] National Institute of Chemistry, Ljubljana, Slovenia

Research Summary

In this book, the results of the chemical modification of cellulose fibres were presented, aimed at protecting the textile material against biodegradation. Namely, cellulose fibres are highly susceptible to microbial attack, resulting in worsened technological and applicable properties of textile products. This is a particularly crucial problem for textiles that are in use. The rate and degree of cellulose biodegradation is affected by several factors, among which the most important are the genera of microorganisms and the environmental conditions needed for microbial growth.

In: Biochemistry Researcher Biographical Sketches ... ISBN: 978-1-62081-308-9
Editors: J. T. Walker and C. M. Jones

Chapter 40

Biogenic Amines: Pharmacological, Neurochemical and Molecular Aspects in the CNS

Tahira Farooqui[1] and Akhlaq A. Farooqui[2]
[1]Aronoff Laboratory, Ohio State Univ., Columbus OH, US
[2]Ohio State University, Columbus, OH, US

Research Summary

Biogenic amines are naturally occurring amines that are derived by enzymic decarboxylation of the natural amino acids. They belong to a class of neurotransmitters including catecholamines (dopamine, norepinephrine, and epinephrine), indolamine (serotonin), and imidazoleamine (histamine). Biogenic amines have great pharmacological and physiological importance. The main objective of this book is to present readers with comprehensive information on pharmacology, neurochemistry and molecular neurobiology of biogenic amine in the CNS of vertebrate and invertebrates in a single volume text. The book has been organized into chapters and sections to provide a better flow of information. Furthermore, this unique volume provides its readers with cutting edge information on biogenic amines. It can be used by graduate students, postdoctoral fellows, researchers, and scientists who work at the pharmaceutical industry as a handbook, which describes all aspects of biogenic amine metabolism.

In: Biochemistry Researcher Biographical Sketches …
Editors: J. T. Walker and C. M. Jones
ISBN: 978-1-62081-308-9

Chapter 41

Biophysics of DNA-Antibiotic Complexes

Nikolai Vekshin
Russian Academy of Sciences, Moscow, Russia

Research Summary

This book presents the basic results on studies of the interaction of anti-cancer Actinomycin antibiotics with DNA. It focuses on interactions of heterocyclic anti-tumor antibiotics (Actinomycins, as typical example) with DNA, poly-nucleotides, oligonucleotides, and aggregates of purines, using spectroscopic methods. Experimental data, various models of structures of the complexes and their physical and chemical properties are described, and possible approaches for delivery of heterocyclic antibiotics to DNA are also discussed.

In: Biochemistry Researcher Biographical Sketches ... ISBN: 978-1-62081-308-9
Editors: J. T. Walker and C. M. Jones

Chapter 42

Borates: Chemical, Pharmaceutical and Pharmacological Aspects

Iqbal Ahmad[1], Sofia Ahmed[1], Muhammad Ali Sheraz[1] and Faiyaz H. M. Vaid[2]
[1]Baqai Medical University, Karachi, Pakistan
[2]University of Karachi, Karachi, Pakistan

Research Summary

Borates are among the most widely used compounds in chemical and related fields. They have extensive applications in consumer products including pharmaceuticals, cosmetics, insecticides, preservatives, detergents, bleaching agents and enamels. This new book provides a thorough background of the chemical, pharmaceutical and pharmacological aspects of borates and their industrial applications. It contains extensive information for academic and industrial research workers in their fields of interest and gives them a collective view of the importance and current state of knowledge in this vastly growing subject.

In: Biochemistry Researcher Biographical Sketches … ISBN: 978-1-62081-308-9
Editors: J. T. Walker and C. M. Jones

Chapter 43

Calixarene Complexes with Solvent Molecules

O. V. Surov, M. I. Voronova and A. G. Zakharov
Russian Academy of Sciences, Ivanovo, Russia

Research Summary

Solid calixarenes as receptors are characterized by phase transitions, which occur in binding of neutral molecules of guest compounds with the formation of clathrates or guest–host intercalation compounds. The process of clathrate formation in the binding of a guest by a host is accompanied by a significant rearrangement of the packing of host molecules, and the result of this rearrangement can depend considerably on the molecular structure of the guest. This new book presents and discusses research on the subject of solid-state complexation.

In: Biochemistry Researcher Biographical Sketches ... ISBN: 978-1-62081-308-9
Editors: J. T. Walker and C. M. Jones

Chapter 44

Chemical and Biochemical Kinetics: New Perspectives

Gennady Zaikov
Russian Academy of Sciences, Moscow, Russia

Research Summary

This book examines the new steps of success in the field of chemical and biochemical kinetics. Topics discussed include the activation energy of molecule and atom diffusion in bio-systems; mathematical modeling dependence of thermodynamic characteristics upon spatial-energy parameter of free atoms; dependence of nanostructure formation degree upon spatial-energy characteristics of free atoms; sevilene-based hybrid composites; organosilicon polymers with photo switchable fragments in the side chain and ferrocene-containing polyphenylenes as precursors for magnetic nanomaterials.

In: Biochemistry Researcher Biographical Sketches ...
Editors: J. T. Walker and C. M. Jones
ISBN: 978-1-62081-308-9

Chapter 45

Chemical and Biochemical Reactions: Kinetics and Mechanism

***G. E. Zaikov*[1], *LinShu Liu*[2], *Rustam Yabukovich Deberdeev*[3] *and O. V. Stoyanov*[3]**

[1]Institute of Biochemical Physics, Moscow, Russia
[2]United States Department of Agriculture, Agricultural Research Service, Wyndmoor, PA, US
[3]Kazan State Technological University, Ka, Russia

Research Summary

This book presents and discusses research in the study of chemical and biochemical kinetics – the science which deals with the changes over time. Topics discussed include the mechanism of catalysis with nickel (II) complexes in toluene and cumene oxidations by dioxygen; the effect of two oxazoles on membrane lipid structure of liver microsomes calculations of bond energy and the structure and properties of polymer adsorption layers on inorganic pigments surface in aqueous dispersion formed under ultrasonic treatment.

In: Biochemistry Researcher Biographical Sketches ... ISBN: 978-1-62081-308-9
Editors: J. T. Walker and C. M. Jones

Chapter 46

Chemical Reactions in Gas, Liquid and Solid Phases: Synthesis, Properties and Application

***G. E. Zaikov*[1] *and R. M. Kozlowski*[2]**
[1]Institute of Biomedical Physics, Moscow, Russia
[2]ESCORENA Program, Poznan, Poland

Research Summary

This new book examines topical data on the subject of chemical reactions in different phases of gas, liquid and solid states, such as the classification of polymers in reactivity toward nitrogen oxide; influence of the initiation rate of radicals on the kinetic characteristics of quercetin and dihydroquercetin in the methyl oleate oxidation; supercritical carbon dioxide swelling of polyheteroarylenes synthesized in N-methylpyrrolidone; inhibition of 2-hexenal oxidation by essential oils of ginger, marjoram, juniper berry, black and white pepper; specific properties of some biological composite materials; properties and applications of aminoxyl radicals in polymer chemistry and others.

In: Biochemistry Researcher Biographical Sketches … ISBN: 978-1-62081-308-9
Editors: J. T. Walker and C. M. Jones

Chapter 47

Chemicals and Radioactive Materials and Human Development

***Rajiv K. Sinha*[1]*, Rohit Sinha*[2] *and Shanu Sinha*[3]**
[1]Griffith University, Brisbane, Australia
[2]Queensland Institute of Medical Research, Brisbane, Australia
[3]Wesley Hospital, Brisbane, Australia

Research Summary

Hazardous chemicals and radioactive substances have permeated the human environment and the ecosystem. They are now in our basic life support systems - air, water, food and soil. A matter of more serious concern is that human beings have been exposed to synthetic chemicals for which there has been no past evolutionary adaptation or experience. Any chemical which interferes directly or indirectly with hormone function can scramble vital messages, derail development and undermine health. This book highlights the scientific research being done to identify these harmful chemicals, especially carcinogens, teratogens, immunotoxins and endocrine disruptors used in production processes and their release into the human environment, making it possible to restrict or even eliminate their production and use.

In: Biochemistry Researcher Biographical Sketches ...
Editors: J. T. Walker and C. M. Jones
ISBN: 978-1-62081-308-9

Chapter 48

Chitosan-Based Hydrogels for Tissue Engineering Applications

G. Luna-Bárcenas, E. Prokhorov, E. Elizalde-Peña, A. Nuno-Licona, I. C. Sanchez, J. E. Gough, C. Velasquillo-Martinez and C. E. Schmidt
Nacional Libramiento Norponiente, Querétaro, México

Research Summary

The polyelectrolyte complexes formed due to electrostatic attraction between the polymers, have improved properties when compared to hybrid CTS-g-GMA, which provides a wide range of applications in the biomedical field. All materials have been characterized by different analytical techniques such as infrared spectroscopy (FTIR), X-ray diffraction (XRD), thermal analysis (DSC and TGA), and the results were compared to the precursor materials (chitosan, X, HA, and CTS-g-GMA). Due to the HA nature, the film obtained from this reaction has been assessed for use as a patch for wound healing; whereas the properties showed by the hydrogels obtained from the reaction with X, make them very promising for applications in the treatment of recovery of spinal cord injuries. This book presents results which strongly support the future use of hydrogels as scaffolds for tissue engineering and recovering.

In: Biochemistry Researcher Biographical Sketches … ISBN: 978-1-62081-308-9
Editors: J. T. Walker and C. M. Jones

Chapter 49

Computational Metabolomics

Nabil Semmar
ISSBAT: Institut Susperieur des Sciences Biologiques Appliquees de Tunis, Tunisia

Research Summary

Metabolism represents a complex system characterized by a high variability in metabolites' structure, concentration and regulation ratio. Such a variability is observed at different metabolic scales going from metabolites to metabolic profiles via chemical reactions and metabolic pathways, as well as under static or dynamic aspects. Variations in these components are due to apparition-disappearance, level increase-decrease and/or relative changes in weights or contributions leading to different structural, functional and evolutive states of the metabolic system. This book presents a variety of different computational approaches of the variability in metabolic systems.

In: Biochemistry Researcher Biographical Sketches ... ISBN: 978-1-62081-308-9
Editors: J. T. Walker and C. M. Jones

Chapter 50

Copolymers in the Preparation of Parenteral Drug Delivery Systems

Rossella Dorati, Claudia Colonna, Ida Genta, Tiziana Modena and Bice Conti
University of Pavia, Pavia, Italy

Research Summary

Poly orthoesters have been the most intensively studied polymers for biomedical applications for their low immunogenicity, good biocompatibility and excellent mechanical properties. In the last decade, the synthesis of polyester-polyether block copolymers has attracted increasing attention, because they may be used in several pharmaceutical and medical applications such as sutures, implants for bone fixation, carriers in drug delivery and temporary matrix or scaffolds in tissue engineering. This book includes the last two decades when block copolymers containing PEO, hydrophilic B-blocks, and PLA or PLGA, hydrophobic A-blocks, have been widely utilized to formulate parenteral drug delivery systems for peptides and proteins, such as micro and nanoparticles, implants, micelles, hydrogels and scaffolds.

In: Biochemistry Researcher Biographical Sketches ... ISBN: 978-1-62081-308-9
Editors: J. T. Walker and C. M. Jones

Chapter 51

Covalently and Noncovalently Bound Assemblies of Calixarenes

M. Deska and W. Sliwa
Jan Dlugosz University of Czestochowa, Poland

Research Summary

Calixarenes, along with calixcrowns are valuable receptors are used in many areas. An advantageous feature of calixarenes is their easy synthesis and functionalization. The modification of calixarene structure offers a variety of new interesting tailor-made species, promising for a variety of applications; one should mention here possibilities of chiral recognition. Calixarenes and calixarene assemblies receive a growing attention today due to their properties allowing applications in various fields. This book is divided into two chapters dealing with covalently and noncovalently bound calixarene assemblies.

In: Biochemistry Researcher Biographical Sketches … ISBN: 978-1-62081-308-9
Editors: J. T. Walker and C. M. Jones

Chapter 52

Current Status of Chitosan on Dermal/Transdermal Drug Delivery Systems

Ipek Ozcan, Taner Senyigit, Evren Homan Gokce and Ozgen Ozer
Ege University, Izmir, Turkey

Research Summary

In case of targeting the drug to the desired part of the skin, vehicles play an important role beside the characteristics of the drug. Many natural and synthetic vehicles have been used for various topical dermal/transdermal preparations. However, chitosan has been standing out with its many advantages based mainly on its biological and physicochemical properties. Chitosan is a unique hydrophilic biopolymer obtained by partial deacetylation of chitin, which is one of the most abundant polysaccharides. It is a natural product widely found in crustacean shells, fungal cell walls, insect exoskeletons, and mollusks. Chitosan is a linear glycosaminoglycan made up of N-acetyl-D-glucosamine units. This book focuses on the use of chitosan for dermal/transdermal drug delivery systems following a general overview of chitosan.

In: Biochemistry Researcher Biographical Sketches … ISBN: 978-1-62081-308-9
Editors: J. T. Walker and C. M. Jones

Chapter 53

Decreasing Oxidative Stress and Retarding the Aging Process

Borut Poljsak
University of Ljubljana, Slovenia

Research Summary

Aging is an inevitable biological process that affects most living organisms. The link between metabolic rate and reactive oxygen species production is an important and long-standing question, and a source of much controversy. A by-product of cell respiration in mitochondria is the formation of reactive oxygen species due to electron leakage from the electron transport chain during oxidative phosphorylation. In simple terms, humans are aging due to oxygen consumption. Damage induced by oxygen appears to be the major contributor to aging and the degenerative diseases of aging such as cancer, cardiovascular disease, immune system decline, and brain dysfunction. This book presents the reasons for oxidative stress formation and the answer to why during the course of evolution the process of free radical damage and defense did not become more perfect so as to produce less free radicals.

In: Biochemistry Researcher Biographical Sketches ...
Editors: J. T. Walker and C. M. Jones
ISBN: 978-1-62081-308-9

Chapter 54

Dipicolinic Acid, its Analogues and Derivatives: Aspects of Their Coordination Chemistry

Alvin A. Holder[1], Lesley C. Lewis-Alleyne[1], Don van Derveer[2] and Marvadeen Singh-Wilmot[2]
[1]University of Southern Mississippi, Hattiesburg, MS, US
[2]Clemson University, Clemson, SC, US

Research Summary

2,6-Pyridinedicarboxylic acid (dipicolinic acid) is a widely used building block in coordination and supramolecular chemistry. The author of this book discusses the coordination chemistry of several metal complexes with dipicolinic acid, its analogues, and derivatives as ligands.

In: Biochemistry Researcher Biographical Sketches ... ISBN: 978-1-62081-308-9
Editors: J. T. Walker and C. M. Jones

Chapter 55

Drug Targeting by Magnetically Responsive Colloids

José L. Arias
University of Granada, Granada, Spain

Research Summary

Magnetic colloids are very promising stimuli-sensitive materials for drug delivery because of their magnetic field responsiveness. An applied magnetic field can concentrate the magnetic nanodevice at the desired site, keeping it there until the drug is totally released. Furthermore, the versatility of magnetic colloids allows the design of multifunctional nanoplatforms with combined therapy and imaging functionalities against diseases with high morbidity. This new book summarizes the most important advances in the formulation and functionalization of such magnetic micro- and nano-particles, and also describes the recent advances on their use in biomedicine.

In: Biochemistry Researcher Biographical Sketches ... ISBN: 978-1-62081-308-9
Editors: J. T. Walker and C. M. Jones

Chapter 56

Electrochemical Technology Applied in Treatment of Wastewater and Ground Water

Chuanping Feng, Miao Li, Xu Guo, Chao Zhao, Zhenya Zhang and Norio Sugiura
China University of Geosciences (Beijing), China

Research Summary

With a rapidly growing economy and burgeoning populations, the world's scarce water resources, such as rivers and lakes, are seriously affected by pollution from the vast discharges of industrial and domestic wastewater, indiscriminate solid waste disposal and runoff from an agricultural sector characterized by excessive use of fertilizer and pesticides and large-scale livestock breeding. Therefore, these wastewaters must be treated before being discharged.

The electrochemical method is an attractive alternative to other methods of wastewater treatment because of its environmental respectability and ease of operation. This new book introduces the application of the electrochemical method for treatment of domestic and industrial wastewater.

In: Biochemistry Researcher Biographical Sketches …
Editors: J. T. Walker and C. M. Jones
ISBN: 978-1-62081-308-9

Chapter 57

Electrospinning Process and Nanofiber Research

***A. K. Haghi*[1] *and G. E. Zaikov*[2]**
[1]University of Guilan, Rasht, Iran
[2]Russian Academy of Sciences, Moscow, Russia

Research Summary

Electrospinning was developed from electrostatic spraying and now represents an attractive approach for polymer biomaterials processing, with the opportunity for control over morphology, porosity and composition using simple equipment. Since electrospinning is one of the few techniques to prepare long fibres of nano- to micrometre diameter, great progress has been made in recent years. This new book presents and discusses research in the study of electrospinning and nanofibers.

In: Biochemistry Researcher Biographical Sketches … ISBN: 978-1-62081-308-9
Editors: J. T. Walker and C. M. Jones

Chapter 58

Folding/Unfolding Kinetics of Lattice Proteins by Applying a Simple Statistical Mechanical Model for Protein Folding

Hiroshi Wako and Haruo Abe
Waseda University, Tokyo, Japan

Research Summary

The folding/unfolding kinetics of a three-dimensional lattice protein was studied using a simple statistical mechanical model for protein folding that was previously developed. The model considers the specificity of an amino acid sequence and the native structure of a given protein. The characteristic relaxation rate on the free energy surface was calculated starting from a completely unfolded structure (or native structure) that is assumed to associate with a folding rate (or an unfolding rate). To elucidate the roles of individual amino acid residues in protein folding/unfolding kinetics, the kinetic properties for all possible single amino acid substitutions of these proteins were calculated and their responses were examined. This book presents and discusses research results in the kinetics of protein folding/unfolding.

In: Biochemistry Researcher Biographical Sketches ... ISBN: 978-1-62081-308-9
Editors: J. T. Walker and C. M. Jones

Chapter 59

Interpolyelectrolyte Complexes of Chitosan

N. A. Samoilova, M. A. Krayukhina and I. A. Yamskov
Russian Academy of Sciences, Moscow, Russia

Research Summary

The results of studies of interpolyelectrolyte complexes of chitosan and different polyanions are summarised and described systematically in this book. Specific properties of chitosan as polyelectrolyte are described. The general concept of the formation of polyelectrolyte complexes is developed. The data on the application of polyelectrolyte complexes of chitosan in medicine and biotechnology, particularly for creation of hemocompatible, thromboresistant materials, bioconstruction materials for replacement of coverlets, blood vessels, bone tissue, immobilization of biological active compounds, and the non-viral vectors of genetic information transfer are discussed herein. This book investigates polyelectrolyte complexes of chitosan.

In: Biochemistry Researcher Biographical Sketches ...
Editors: J. T. Walker and C. M. Jones

ISBN: 978-1-62081-308-9

Chapter 60

Modern Nanochemistry

A. K. Haghi[1] and G. E. Zaikov[2]
[1]University of Guilan, Iran
[2]Russian Academy of Sciences, Russia

Research Summary

Nanochemistry is the use of synthetic chemistry to make nanoscale building blocks of desired shape, size, composition and surface structure, charge and functionality with an optional target to control self-assembly of these building blocks at various scale-lengths. This new book traces the fascinating modern practice of the chemistry of the non-covalent bond from its fundamental origins through to its expression in the emergence of nanochemistry. Fusing synthetic materials and supramolecular chemistry with crystal engineering and the emerging principles of nanotechnology, this book is an ideal introduction to current chemical thought for researchers and a superb resource for students entering these exciting areas for the first time.

In: Biochemistry Researcher Biographical Sketches ... ISBN: 978-1-62081-308-9
Editors: J. T. Walker and C. M. Jones

Chapter 61

Polymer Assisted Surface Modification by Photons

Eduardo Weibel
Instituto de Química, UFRGS, Brasil

Research Summary

In recent years, many advances have been made in developing surface treatments to alter the chemical and physical properties of polymer surfaces without affecting their excellent bulk properties. In particular, the past two decades have seen the rapid development of new strategies for the design of polymer functional surfaces. As one of the main techniques developed to achieve efficient surface modification of polymeric materials, electromagnetic radiation-induced surface modification has been widely applied as simple, useful and versatile approach to improve the surface properties of polymers. This new book examines the recent advances in electromagnetic radiation–induced surface modification of polymers, predominantly focusing on the research work carried out in the last decade.

In: Biochemistry Researcher Biographical Sketches … ISBN: 978-1-62081-308-9
Editors: J. T. Walker and C. M. Jones

Chapter 62

Porous Structure and Adsorption Behaviours of Chitosan

O. K. Krasilnikova, O. V. Solovtsova, E. V. Khozina, T. Y. Grankina, N. V. Serebryakova and S. M. Shinkarev
A. N. Frumkin Institute of Physical Chemistry and Electrochemistry RAS, Moscow, Russia

Research Summary

This book reviews the interrelation between the structure and adsorption properties of water and nitrogen vapors and Ni, Cu and Ag cations of the chitosan samples modified by freeze-drying after precipitation with NaOH or Na2CO3 or drying in air, as well as cross-linking. The investigations by IR spectroscopy, X-ray diffraction and pulsed NMR show the influence of the preliminary treatments on the supramolecular structure. The more ordered structure is observed in the freeze-dried samples of chitosan compared with the air-dried sample. The parameters of porous structure of the chitosan samples calculated from the isotherms of water vapors adsorption and the nitrogen vapors low temperature adsorption are compared. It is revealed that the isotherms of water vapors adsorption on the chitosan samples are more sensitive to the structural modification of polymers.

In: Biochemistry Researcher Biographical Sketches …
Editors: J. T. Walker and C. M. Jones
ISBN: 978-1-62081-308-9

Chapter 63

Principles of Free Radical Biomedicine. Volume I

Kostas Pantopoulos and Hyman M. Schipper
Lady Davis Institute for Medical Research, Montreal, QC, Canada

Research Summary

Recent years have witnessed an avalanche of new knowledge implicating free radicals in virtually every aspect of biology and medicine. It is now axiomatic that the regulated accumulation of reactive oxygen species (ROS) contributes to organismal health and well-being and that ROS serve as signaling molecules involved in cell growth, differentiation, gene regulation, replicative senescence and apoptosis. This book is an interdisciplinary text broken up into three consecutive volumes on the biochemistry and cellular/molecular biology of free radicals, transition metals, oxidants and antioxidants, and the role of oxidative stress in human health and disease.

In: Biochemistry Researcher Biographical Sketches ... ISBN: 978-1-62081-308-9
Editors: J. T. Walker and C. M. Jones

Chapter 64

Protein-Protein Interactions

Pandjassarame Kangueane
Bioinformatics & Biotechnology, AIMST, Malaysia

Research Summary

This book describes the fundamental physical, chemical, molecular and biological aspects of proteins binding to one another in cellular biology. PPI is a common phenomenon in cell and molecular biology events. The networks of PPI describe cellular function in both unicellular and multi-cellular organisms. Homo complexes (dimers of identical monomers) or hetero complexes (dimers of non-identical monomers) are involved in PPI. These complexes are common in catalysis, regulation and structural assembly. However, the molecular principles of protein interactions are difficult to understand due to the geometrical and chemical characteristics of proteins. The challenge today is to identify interaction partners and sites from sequence and/or structures. This book describes the principles (physics and chemistry) of PPI and protein interfaces to develop models for PPI predictions.

In: Biochemistry Researcher Biographical Sketches ...
Editors: J. T. Walker and C. M. Jones
ISBN: 978-1-62081-308-9

Chapter 65

Recent Developments in Bio-Nanocomposites for Biomedical Applications

Ashutosh Tiwari
National Institute for Materials Science, Tsukuba, Japan

Research Summary

Bio-nanocomposites form a unique class of a research area that integrates biology, chemistry, materials science, engineering and nanotechnology to present an interdisciplinary approach for solving of problems. In today's world, bio-nanocomposites are becoming increasingly prevalent due to the extraordinary properties that they possess. Scientists learn to select suitable matrices (e.g. aliphatic polyesters, polypeptides and proteins, polysaccharides, and polynucleic acids) and fillers (e.g. nanotubes, nanofibers, clay nanoparticles, hydroxyapetite and metal nanoparticles) and alter their chemistry and structure to suit the target field. This book provides the most recent research made in the field of bio-nanocomposites as applied to biomedical fields including drug-delivery, biosensors, cancer diagnosis, and tissue engineering.

In: Biochemistry Researcher Biographical Sketches ... ISBN: 978-1-62081-308-9
Editors: J. T. Walker and C. M. Jones

Chapter 66

Sampling and Sample Preparation in Analytical Chemistry

Jaroslava Švarc-Gajic
Bulevar cara Lazara 1, Faculty of Technology, Serbia

Research Summary

The art of sample manipulation considers adequate sample collection, preservation and storage, as well as its safe preparation. Analytical chemists often find themselves in a position to spend considerable more time and effort in preparing samples rather than analyzing them. In many cases the reliability of analysis is limited by the correctness of sample manipulation. The skills of sample manipulation are based on abundant knowledge, competence and rich experience. Being indispensable in every laboratory, educational or scientific institution, this book encompasses all relevant issues related to collecting samples of any kind, as well as methods of sample preparation for specific analytical goal. The presented material provides an overview of sampling principles and strategies supporting them with inevitable statistical background.

In: Biochemistry Researcher Biographical Sketches … ISBN: 978-1-62081-308-9
Editors: J. T. Walker and C. M. Jones

Chapter 67

Soybeans: A Peroxidase Source for the Biotreatment of Effluents

J. L. Gomez, M. Gomez and M. D. Murcia
Murcia University, Murcia, Spain

Research Summary

Among the numerous applications of soybean, its use as a source of soybean peroxidase is of particular interest in the area of bioremediation. Peroxidases are oxidoreductase enzymes with a wide variety of substrate specificity, that catalyze, in the presence of hydrogen peroxide, the oxidative polymerization of different pollutants widely found in industrial effluents. These pollutants can be highly toxic for both human beings and the environment and they must be removed from wastewater before they are discharged into the environment. The use of enzymes is presented as an alternative method for the industrial wastewater treatment when conventional physical, chemical and biological methods such as adsorption in carbon active, chemical oxidation and microorganism treatment may be ineffective.

This book discusses the various advantages of soybeans as a source for the biotreatment of hazardous effluents.

In: Biochemistry Researcher Biographical Sketches … ISBN: 978-1-62081-308-9
Editors: J. T. Walker and C. M. Jones

Chapter 68

Squalene: Current Knowledge and Potential Therapeutical Uses

A. Ramírez-Torres, C. Gabás and C. Barranquero
Universidad de Zaragoza, Salud del Gobierno de Aragón, Spain

Research Summary

Squalene owns a long tradition of research since its first publication backs to 1926. To provide a state of the art revision, this book has collected published works regarding: in vivo and in vitro antioxidant properties, the bioavailability in animal models and in humans, the plasma and intracellular transport of this lipophilic molecule, its ability to neutralize different xenobiotics, its anti-inflammatory, anti-atherosclerotic and anti-neoplastic properties, its presence in skin lipids and its role in skin aging and pathology, and finally its use to boost antigenicity of vaccines.

In: Biochemistry Researcher Biographical Sketches ... ISBN: 978-1-62081-308-9
Editors: J. T. Walker and C. M. Jones

Chapter 69

Starch Science Progress

***Luybov A. Wasserman*[1], *Gennady E. Zaikov*[1], *Piotr Tomasik*[2] and *Robert G. Gilbert*[3]**
[1]Institute of Biochemical Physics RAS, Moscow, Russia
[2]Agricultural University of Cracow, Poland
[3]Centre for Nutrition and Food Sciences, Queensland, Australia

Research Summary

Polysaccharides, particularly starches, are common all over world. They are considered an importqant energy source and substitute of fossils including petrol because of their availability, biodegradability and versatility. Such anticipations generate intensive studies on structure, properties and novel ways of using polysaccharides- and starch-based materials. This book contains original reviews and research papers presented at the XVII International Starch Convention, Moscow-Cracow, which was held in Moscow in June, 2009.

In: Biochemistry Researcher Biographical Sketches ... ISBN: 978-1-62081-308-9
Editors: J. T. Walker and C. M. Jones

Chapter 70

Symbiotic Biofilms and Brain Neurochemistry

Alexander V. Oleskin, Vladimir I. Shishov and Kristina Malikina
Moscow State University, Moscow, Russia

Research Summary

An overwhelming majority of known species of microorganisms form biofilms, i.e. spatially and metabolically structured communities embedded in the extracellular biopolymer matrix. Biofilm development is a complex multi-stage process involving reversible and, at a later stage, irreversible attachment of microbial cells to the substrate surface. Development also involves matrix formation, three-dimensional structuring of the whole community including the formation of mushroom- or pillar-shaped structures and, finally, the degradation of the biofilm and the dispersion of the cells involved. This book examines these processes in the example of microorganisms that interact with the animal or human organisms, playing the roles of symbionts or pathogens.

In: Biochemistry Researcher Biographical Sketches … ISBN: 978-1-62081-308-9
Editors: J. T. Walker and C. M. Jones

Chapter 71

Tannins and Phenolics in Animal Nutrition: Chemistry and Challenges

Miguel E. Alonso-Amelot
University of Los Andes, Venezuela

Research Summary

Feeding the world's human population of nearly seven billion people is probably the most demanding challenge we face in the 21st century. We need to serve ourselves between 2500 and 3200 kcal of energy from food every day to survive and perform our daily chores. About a third of this figure comes from animal products. Traditionally, tannins in forage, browse, silage and grains for feed formulations have been thought of as classical antinutritionals. Today, however, this concept is changing as our knowledge about the intimacies of tannin-food-digestion chemistry and gut bacterial interaction improves. The central issue of this book is to examine how far these developments around tannins go in the area of nutrition and the chemistry involved, and the manner in which in vitro and in vivo models illustrate their beneficial as well as troublesome aspects.

In: Biochemistry Researcher Biographical Sketches … ISBN: 978-1-62081-308-9
Editors: J. T. Walker and C. M. Jones

Chapter 72

Theoretical Estimation of Acidic Force of Linear Olefins of Cationic Polymerization

G. E. Zaikov[1] and Babkin Vladimir Alexandrovich[2]
[1]Russian Academy of Sciences, Moscow, Russia
[2]Volgograd State Architect-Building University, Volgograd, Mikhailovka

Research Summary

This new book presents research results on the quantum-chemical calculation of linear olefins of cationic polymerization (a geometrical and electronic structure, the general and electronic energy, distribution of charges to atoms and so forth). In addition, acid force of these monomers is theorized. The book is intended for science officers, competitors, post-graduate students and persona working for a doctorate degree in the field of polymer chemistry.

In: Biochemistry Researcher Biographical Sketches ...
Editors: J. T. Walker and C. M. Jones
ISBN: 978-1-62081-308-9

Chapter 73

Alkaloids: Properties, Application and Pharmacological Effects

Monika Mahajan, Vinay Kumar and Sudesh Kumar Yadav*
Biotechnology Division, Institute of Himalayan Bioresource Technology, CSIR, HP, India

Research Summary

Alkaloids are one of the most abundant and diverse group of secondary metabolites found at 10-15% concentration in almost all plants. Alkaloids are heterocyclic compound containing nitrogen in a negative oxidation state due to which they can form salt with acids making them water soluble. Most of them are colorless, crystalline solid, bitter in taste and having a definite melting point. These are classified on the basis of carbon skeletal structure and biogenic origin. Several of them have been known for mankind for several thousand years. Alkaloids disrupt the integrity of biomembrane, potent inhibitors of ion channel and impair the function of microtubules or microfilaments. Alkaloids are mutagenic and carcinogenic due to their properties to bind DNA and acts at DNA and RNA polymerase level. A list of alkaloids acts as antibiotics and inhibits the translation process in various organisms. Alkaloids either activate or inhibit the central process at cellular and organ level in animals. They affect the digestive process by interfering the hydrolytic enzymes. They impaired the kidney and liver functioning. Some alkaloids influence the reproductive system of animals. Alkaloids also possess antimitotic and allergic effects at cellular level. These functions of alkaloids are well known for their toxic and sometimes psychomimmetic, euphoric, and hallucinogenic properties. Alkaloids also have role in chemical ecological perspectives like plant-herbivore interaction, plant-plant interaction and plant-microbial interaction. Many alkaloids, though poisonous, have physiological effect that renders them valuable medicine

* Correspondence: skyt@rediffmail.com; sudeshkumar@ihbt.res.in All authors contributed equally to this manuscript.

against various diseases including malaria, diabetics, cancer, cardiac dysfunction etc. These are also used in local anesthesia and relief of pain. This article presents a thorough understanding of alkaloids with respect to their properties, applications and pharmacological effects.

In: Biochemistry Researcher Biographical Sketches ... ISBN: 978-1-62081-308-9
Editors: J. T. Walker and C. M. Jones

Chapter 74

Intraocular Pressure Lowering Effect of Natural Ergot Alkaloids and their Future Applications in Ocular Pharmacology

Gustavo Puras[*]
Networking Biomedical Research Center on Bioengineering, Biomaterials and Nanomedicine, CIBER-BBN, SLFPB-EHU, Spain

Research Summary

The ergot alkaloids are among the most important natural pharmaceuticals and toxins in human history. The word ¨Ergot¨ refers to a fungal infection caused by *Claviceps purpurea* which infects grains of rye and related grasses. The dark sclerotia formed in the final disease stages is the source of the pharmaceutically interesting ergot alkaloids. Through history some ergot alkaloids have been ingested accidentally leading to poisoning, some have been used illicitly for psychedelic recreation and some have been used for medical benefit over the ages. Naturally occurring ergot consists of two types of alkaloids. The first type derives from 6,8-dimethylergoline and the second one consists of the peptide alkaloids or lysergic acid derivatives which are the pharmacologically active alkaloids.

The complex pharmacological profile of natural ergot alkaloids is mostly due to the structural similarity between a lysergic acid derived compound and neurotransmitters like noradrenalin, dopamine and serotonin. This broad specificity can provoke undesirable side effects. However, a greater understanding of natural ergot alkaloids has been developed with modern research, and their utilization as pharmaceuticals has become more refined and effective. One of the most promising approaches to improve natural ergot alkaloids as therapeutic agents involves narrowing the specificity of the compounds by chemical modifications in order to retain the therapeutic effects and avoid adverse effects.

[*] Email:gustavopuras@yahoo.es

In the ocular pharmacology field, ergot derivatives were suggested at an early stage as a potent group of anti glaucoma drugs. Glaucoma is a major sight threatening optic neuropathy that is the second cause of blindness in the developed countries. An important risk factor associated with glaucoma is the elevated intraocular pressure. At the moment many classes of drugs used clinically are known to reduce intraocular pressure; however there is a need for more efficacious agents that provide long duration of intraocular pressure reduction and fewer undesirable side-effects.

In this chapter we describe the preclinical powerful intraocular pressure-lowering effect of topically applied natural ergot alkaloids ergocristine, α-ergocryptine and ergocornine in ocular normotensive and hypertensive rabbits. Our results suggest that these compounds and related drugs could have a place in anti glaucoma therapy. However, further studies are needed to elucidate the intimate mechanisms of their ocular hypotensive effect, since natural ergot alkaloids have a complex pharmacological profile.

In: Biochemistry Researcher Biographical Sketches ...
Editors: J. T. Walker and C. M. Jones
ISBN: 978-1-62081-308-9

Chapter 75

Pilocarpine and Related Alkaloids in *Pilocarpus* Vahl (Rutaceae)

***Alexandra C. H. F. Sawaya*[1], *Ilka N. Abreu*[1,2], *Nathalia L. Andreazza*[1], *Marcos N. Eberlin*[3] and *Paulo Mazzafera*[1,*]**

[1]Plant Biology Department, Institute of Biology,
State University of Campinas, SP, Brazil
[2]Scottish Crop Research Institute,
Department of Plant Products and Food Quality,
Invergowrie, Scotland, UK
[3]Thomson Mass Spectrometry Laboratory,
Institute of Chemistry, State University of Campinas,
UNICAMP, SP, Brazil

Research Summary

Pilocarpine is mainly known as a drug for the treatment of glaucoma and it is also used as a stimulant of sweat and lachrymal glands. Species of the genus *Pilocarpus* are collectively named *jaborandi* in Brazil and their leaves are the only known source of this imidazole alkaloid. Pilocarpine is mainly obtained from two species, *Pilocarpus microphyllus* and *Pilocarpus jaborandi* and, despite the economical and pharmacological importance of this alkaloid, very little is known about pilocarpine, from basic information on the contents in different jaborandi species and plant tissues to the biosynthetic route and the metabolic control. This review will focus briefly on the genus jaborandi, then on what is known about pilocarpine biosynthesis followed by possible biotechnological applications aiming to produce the alkaloid *in vitro*. Finally, the alkaloids found in this genus, their plant sources and pharmacological applications will be reviewed.

[*] Corresponding author: pmazza@unicamp.br.

In: Biochemistry Researcher Biographical Sketches ... ISBN: 978-1-62081-308-9
Editors: J. T. Walker and C. M. Jones

Chapter 76

Quaternary Benzo[c]phenantridine Alkaloids: Perspective Fluorescence DNA probes

Petr Táborský[1]***, Iva Slaninová***[2] ***and Eva Táborská***[3]
[1] Dept. of Chemistry, Faculty of Science, Masaryk University, Brno, Czech Republic
[2] Dept. of Biology, Faculty of Medicine, Masaryk University, Brno, Czech Republic
[3] Dept. of Biochemistry, Faculty of Medicine, Masaryk University, Brno, Czech Republic

Research Summary

Quaternary benzo[c]phenantridine alkaloids (QBAs) belong to the group of isochinoline alkaloids. This sort of alkaloids is naturally occurring mainly in plants from families *Papaveraceae, Fumaraceae, Ranunculaceae* and *Rutuaceae*. At present, at least 16 natural alkaloids of this type are known (Table 1). According to the substitution of benzo[c]phenanthridine ring system two main types of QBAs are distinguished. Nitidine group includes 2, 3, 8, 9-tetrasubstituted alkaloids. The main representatives are nitidine, avicine and fagaronine. For sanguinarine group is typical substitution in positions 2, 3, 7, 8 on the aromatic skeleton. The typical alkaloids of this group are chelerythrine and sanguinarine. As minor alkaloids in some plants species also pentasubstituted derivatives of sanguinarine group sanguilutine, sanguirubine, chelirubine, chelilutine or hexasubstituted macarpine were described. Their biosynthetic precursors 10-hydroxysanguinarine, 10-hydroxychelerythrine and 10-hydroxychelirubine were isolated in experiments with cell cultures.

Except of important biological activities such as antiproliferative, antiapoptotic, antiinflamatory and antimicrobial effects QBAs show also interesting spectral properties. All QBAs are colored (from yellow to red) and exhibit fluorescence after light irradiation in wide region of wavelengths. Moreover, the fluorescence properties are in many cases enhanced (or suppressed) in presence of double stranded DNA [1]. This effect can be employed in DNA quantification by fluorescence spectroscopy and related techniques. In addition to interesting changes in spectral properties, all QBAs quickly permeate to living cells and interact with cellular structures. Moreover sanguinarine, chelirubine, sanguirubine and macarpine interact with DNA. These characteristics predict multiple applications of the above-mentioned QBAs with significant diagnostic and analytical utility. They can be used as supravital fluorescent DNA probes both in fluorescence microscopy and flow cytometry (including multiparameter analysis).

In: Biochemistry Researcher Biographical Sketches ... ISBN: 978-1-62081-308-9
Editors: J. T. Walker and C. M. Jones

Chapter 77

Metabolic Engineering and Alkaloid Producing Pathways: Scopolamine as a Model Case

Sandra Irene Pitta-Alvarez*
Instituto de Ciencia y Tecnología Dr. César Milstein, Fundación Pablo Cassará, Buenos Aires, Argentina

Research Summary

Alkaloids are molecules widely used in medicine. The majority are extracted from plants, although some can be chemically synthesized. Scopolamine is a tropane alkaloid employed in medicine as an anticholinergic, and its properties range from pre-operative medicine to motion sickness. Hairy roots, transformed with *Agrobacterium rhizogenes*, have been found to be suitable for the production of secondary metabolites, such as scopolamine, because they are stable and highly productive in hormone-free culture conditions. There have been several approaches to increase scopolamine accumulation in hairy root cultures, among them elicitation, the use of plant-growth regulators and the modification of culture conditions. However, metabolic engineering techniques currently represent the most promising strategy to achieve competitive yields of this alkaloid.

* spitta@fibertel.com.ar.

In: Biochemistry Researcher Biographical Sketches ... ISBN: 978-1-62081-308-9
Editors: J. T. Walker and C. M. Jones

Chapter 78

The Effects of Alkaloids on the Feeding of Lepidopteran Larvae

Vonnie D. C. Shields*[*] *and Timothy L. Martin
Towson University, Towson, MD, US

Research Summary

Secondary plant compounds, such as alkaloids, are important mediators of insect plant interactions. In this chapter, we describe the role of alkaloid phytochemicals in insect behavior and physiology on the taste system of gypsy moth larvae, *Lymantria dispar* (L.), a highly polyphagous pest species which feeds on the foliage of a few hundred plant species. These larvae possess two pairs of sensory organs on their mouthparts, namely the lateral and medial styloconic sensilla, each bearing four bipolar chemosensory neurons. These organs are thought to be the main sensory organs involved in feeding and play an important role in host-plant selection and the detection of alkaloids. Many alkaloids act as potent feeding deterrents in lepidopteran insects. In order to determine the potency of selected alkaloids as feeding deterrents in gypsy moth larvae, we applied them exogenously to natural (i.e., leaves) versus artificial substrates (i.e., filter paper) to compare their overall effect. In addition, we also carried out dose-response experiments to determine both the deterrency threshold and the approximate concentration that reduced feeding by 50% for these alkaloids. Lastly, we applied selected alkaloids to the tip of the medial styloconic sensillum, found to house a deterrent-sensitive taste receptor cell, using an electrophysiological tip-recording method to characterize the neurophysiological responses of this cell to selected alkaloids. Overall, the results of our studies suggest alkaloids as novel candidates in designing suitable strategies for crop protection from insect pests.

* Corresponding author: Shields, V.D.C. (vshields@towson.edu), Phone: 410-704-3130, Fax: 410-704-2405, E-mail:vshields@towson.edu.

In: Biochemistry Researcher Biographical Sketches ...
Editors: J. T. Walker and C. M. Jones
ISBN: 978-1-62081-308-9

Chapter 79

Biosynthesis of Chlorinated Alkaloids in *Menispermum Dauricum* Root Culture

Yukihiro Sugimoto and Rieko Hori
Graduate School of Agricultural Science, Kobe University
Rokkodai, Nada, Japan

Research Summary

A large number of halogenated organic compounds have been isolated from plants, microorganisms, insects and animals. However, not much is known about the mechanisms by which the halogen atoms are incorporated into organic compounds. Cultured roots of *Menispermum dauricum*, the Asian vine, produce four structurally related chlorinated isoquinoline alkaloids; acutumine (1), acutumidine (2), dauricumine (3) and dauricumidine (4). Additionally *M. dauricum* roots produce their dechloro-derivatives dechloroacutumine (5) and dechlorodauricumine (6). The structures of these alkaloids were established by spectroscopic, crystallographic and chemical methods. These alkaloids were labeled with ^{3}H, ^{13}C, ^{15}N, or ^{36}Cl, isolated, and fed independently to the *Menispermum* root culture. Biosynthetic relationships among the chlorinated alkaloids and their dechloro-derivatives were studied by monitoring incorporation of tracers derived from one alkaloid to the others, using a liquid scintillation counter for the radio isotopes ^{3}H and ^{36}Cl, an NMR spectrometer for ^{13}C, and a mass spectrometer for ^{15}N. Based on the results obtained from these tracer experiments, biosynthetic pathway from tyrosine to the chlorinated alkaloids are proposed, with special attention to steps related to introduction of a chlorine atom into the alkaloid molecules.

In: Biochemistry Researcher Biographical Sketches ... ISBN: 978-1-62081-308-9
Editors: J. T. Walker and C. M. Jones

Chapter 80

Lipids and Cell Function

***Anna Karenina Azevedo-Martins*[1], *Thais Martins de Lima Salgado*[2], *Renata Gorjão*[3], *Érica Portioli Silva*[2], *Jarlei Fiamoncini*[2], *Maria Fernanda Cury-Boaventura*[3], *Elaine Hatanaka*[3] *and Rui Curi*[2]**

[1]School of Arts, Science and Humanities, University of São Paulo, Brazil
[2]Institute of Biomedical Sciences, University of São Paulo, Brazil
[3]University Cruzeiro do Sul, Brazil

Research Summary

This chapter approaches biochemical, physiological and therapeutic aspects of some components of the large class of lipids. We begin with a brief history of essential FA and finish presenting the main techniques used in research to analyse the cell content of fat. Lipids are not only the best form of energy storage, cell membrane constituent, or hormones precursors. Lipids and their metabolites participate in intra and intercellular signalling pathways and regulate gene expression. Recent studies on cytotoxicity of FA are discussed. A significant part of the content herein presented is product of the scientific investigation of the own authors and new results, not yet published, are also shown.

In: Biochemistry Researcher Biographical Sketches ... ISBN: 978-1-62081-308-9
Editors: J. T. Walker and C. M. Jones

Chapter 81

Immune Response Modulation by Targeted Complexes Based on Streptavidin

Zsuzsanna Szekeres[2], Melinda Herbáth and József Prechl[2]
[1]Department of Immunology, ELTE, Pazmany P. s. 1/C Budapest, Hungary
[2]Immunology Research Group, MTA Pazmany P. s. 1/C Budapest, Hungary

Research Summary

Several immunomodulatory constructs based on streptavidin-biotin binding have been described that exploit the modularity and flexibility of this system. In general, a streptavidin core serves as the carrier of targeting and effector molecules. Targeting agents, such as whole antibodies or their fragments and natural ligands, deliver antigen to specific receptors on leukocytes. This targeted delivery can both enhance and fine-tune immune responses, which are directed against streptavidin itself or its cargo. Due to its tetrameric structure, streptavidin-based targeting complexes can crosslink cell surface receptors. This property enables such constructs to initiate signaling events in the target cells. Thus, in addition to improving antigen uptake, activation of APCs can be achieved. By using liposome- or nanobead-bound streptavidin, one can both boost the crosslinking effect and target different cell populations, specialized in the uptake of particulate antigens. To generate streptavidin-based constructs, various molecular biological and chemical methods have been applied. In a preferred setup, monobiotinylated components are used, in order to ensure the formation of complexes with controllable stoichiometry. This strategy also allows the combination of distinct targeting and antigenic moieties in the same complex. Here we review various approaches of streptavidin-based immunotargeting and their effects on immune responses.

In: Biochemistry Researcher Biographical Sketches ...
Editors: J. T. Walker and C. M. Jones
ISBN: 978-1-62081-308-9

Chapter 82

Electrochemical and Optical Biosensors Based on Strept(Avidin)-Biotin Affinity

XinRan Cheng and Kagan Kerman
University of Toronto, Canada

Research Summary

Streptavidin is a ~53kDa tetrameric protein purified from the bacterium *Streptomyces avidinii.* It is well known to contain four identical subunits, each containing a single binding site that binds very tightly to the small molecule biotin, also known as vitamin H. This high affinity ($K_{assoc}= 10^{14}M^{-1}$) has been applied widely in biotechnology and biosensors. This review will cover the latest trends in the application of streptavidin-biotin system in DNA sensors and immunosensors using electrochemical and optical techniques. Streptavidin is sometimes used in connection with magnetic beads in DNA sensing. By immobilizing the biotinylated DNA probes on streptavidin-conjugated magnetic beads, the target DNA can be easily isolated from a complex matrix using an external magnet. Electrochemical and fluorescence-based immunosensors utilize the biotinylated antibodies with streptavidin-conjugated enzymes (horseradish peroxidase, alkaline phosphatase, etc.) or streptavidin-conjugated fluorescent dyes as the source of analytical signals. Surface plasmon resonance (SPR) and Localized SPR, which provide label-free analysis of dynamic surface events, are valuable optical tools for studying interactions between biomolecules. In immunosensors, SPR is typically applied to detect the formation of a sandwich-type immune complex comprised of a capturing antibody immobilized on a sensor surface, the target biomolecule, and a detecting antibody. The well-studied streptavidin-biotin system serves as an excellent model for the development of LSPR biosensors using metal nanoparticles. This review will also describe the recent fundamental spectroscopic studies that reveal key relationships governing the LSPR spectral location and its sensitivity to the local environment, including changes on the substrate due to binding events. The immobilization of streptavidin on gold nanoparticles is detected with the changes in the peak wavelength and intensity of the LSPR absorption band. Further attachment of the biotinylated ligands on streptavidin-conjugated gold nanoparticles can be detected at high sensitivity and selectivity. The results from these biosensor studies guide the design of new sensing experiments, illustrated through applications of nanomaterials, in which researchers use both the streptavidin-biotin affinity and the advanced transduction techniques to detect molecules of chemical and biological relevance.

In: Biochemistry Researcher Biographical Sketches … ISBN: 978-1-62081-308-9
Editors: J. T. Walker and C. M. Jones

Chapter 83

Cardiac (Patho)Physiological Actions of the Classical Mu-, Delta-, and Kappa-Opioid Receptor System

Craig S. Bolte[1,2], *Garrett J. Gross*[3] *and Jo El J. Schultz*[1]
[1]Department of Pharmacology and Cell Biophysics,
University of Cincinnati, US
[2]Department of Neonatology, Division of Pulmonary Biology,
Cincinnati Childrens' Hospital Research Foundation, US
[3]Department of Pharmacology and Toxicology,
Medical College of Wisconsin, US

Research Summary

The endogenous opioidergic system is known to express both receptors and peptides within the cardiovascular system, including within the heart itself (Barron, 2000). Precursors for all three families of endogenous opioid peptides (EOPs), prodynorphin (PDYN), proenkephalin (PPE) and proopiomelanocortin (POMC) are generated in the heart and alternative processing results in a myriad of mature peptides (Barron, 1999). Biosynthesis of the opioid peptide precursors, prodynorphin and proenkephalin, occurs predominantly in ventricular cardiomyocytes, not in the atria (Howells et al., 1986; Springhorn and Claycomb, 1989, 1992; Younes et al., 2000); whereas, processing of proopiomelanocortin to the cardiac β-endorphin peptide occurs in atrial myocytes (Millington et al., 1999). PPE has been shown to be more highly expressed in the adult heart (McLaughlin and Wu, 1998) and higher in the ventricles than in the atria (Barron, 2000; Weil et al., 1998), with left-sided preference (Weil et al., 1998; Weil et al., 2006). Similarly, POMC has been reported to only be detectable in the adult heart (McLaughlin and Wu, 1998) and to be preferentially localized to the ventricles compared to the atria (Millington et al., 1999). However, PDYN is reported to only be detected in the atria (McLaughlin and Wu, 1998). Similar to the biosynthesis of cardiac peptides from their prohormone precursors, cardiac opioid peptides are spatially distributed; levels of enkephalins are higher in ventricular tissue (Barron et al., 1995; Springhorn and

Claycomb, 1992) and β-endorphin peptide is predominantly found in atrial tissue (Millington et al., 1999). In atrial tissue, enkephalins are packaged in vesicles, but are found unpackaged in ventricles (Barron, 1999, 2000; Barron et al., 1995; Millington et al., 1999).

In: Biochemistry Researcher Biographical Sketches ... ISBN: 978-1-62081-308-9
Editors: J. T. Walker and C. M. Jones

Chapter 84

Angiotensin Converting Enzyme Inhibitors: A Class of Potent Antihypertensive Agents

Sharad Kumar Panday*, Jagdish Prasad and Manohar Bhushan Pathak
Department of Applied Chemistry,
Faculty of Engineering and Technology,
M. J. P. Rohilkhand University, Bareilly, U. P., India

Research Summary

Hypertension, a major cause of heart-diseases is caused by malfunction of one or more of the complex set of mechanisms controlling blood pressure. In last four decades specific agents interacting with these regulatory mechanisms have become available. Introduction in early eighties of Captopril, a rationally designed potent inhibitor of Angiotensin converting enzymes (ACE), heralded a new era in antihypertensive therapy. Since then rapid strides have been made in understanding of the active site and the spatial requirements for ACE inhibitors. This chapter has been aimed to report the existing knowledge and to enable researchers all over the world to exploit further with an objective to develop better therapeutic agents.

Bio receptors being chiral, interactions with organic molecules are strongly influenced by the chirality of the substrate. The two enantiomers of a drug are some times known to elicit differential and in some cases even opposite responses, amply heightened by work on ACE inhibitors. The present trend, in drug research, is to design and synthesize molecules with designated chirality. ACE inhibitors represent a class of rationally designed drugs based on the comprehensive knowledge of the enzyme and its binding sites and had led to great revolution in antihypertensive therapy. ACE is a Zn^{++} containing metalloenzyme similar to carboxypeptidases, whose binding sites have been studied properly and a hypothetical model incorporating all the binding sites have been developed. Based on the hypothetical model for

* E-mail: drsharadpandey@yahoo.com.

ACE binding sites and the role which it plays in living organism, attempts have been made to develop potent inhibitors of ACE with an objective to control blood pressure. Early researches on ACE inhibitors started with snake venom peptides, showing antihypertensive activity. Though these could not be used as drugs due to high toxicity but certainly opened an avenue to look for development of antihypertensive agents on these lines. Further researches coupled with structure activity relationship (SAR) studies led to development of various ACE inhibitors and ultimately in early eighty's first rationally designed drug "Captopril" was introduced in market. Since then various other drugs acting through same mechanism and controlling blood pressure effectively have come out.

The chapter entitled "Angiotensin converting enzymes inhibitors: A class of potent antihypertensive agents" is based on literature reports and intended to summarized the available knowledge and to give an update picture of major advances towards the design and synthesis of angiotensin converting enzymes inhibitors and their role in controlling blood pressure and shall undoubtedly be valuable to readers working in the area. This chapter gives information about the designing and developing of ACE inhibitors reported till date with future prospectives to achieve further progress as for as designing and development of even better ACE inhibitors is concerned.

In: Biochemistry Researcher Biographical Sketches ...
Editors: J. T. Walker and C. M. Jones
ISBN: 978-1-62081-308-9

Chapter 85

Protein Superfamilies Based Phylogenomic Analysis of Archaeal Domain

***P. Chellapandi*[1]* *and S. Sivaramakrishnan*[2]**
[1]Department of Bioinformatics, School of Life Sciences, Bharathidasan University, Tiruchirappalli, Tamil Nadu, India
[2]Department of Biotechnology, School of Life Sciences, Bharathidasan University, Tiruchirappalli, Tamil Nadu, India

Research Summary

Since a metabolic pathway with similar function has similar evolutionary relationship, the operation of metabolic pathways is essential to the survival of organisms, and metabolism related genes usually constitute only 10–20% of the total number of genes, metabolic pathway-based analysis can minimize the effect of the gene number, and thus allow more precise investigation of evolutionary relationships. Herein, we have shown that protein superfamilies of archaea were diverged and metabolically converged together within the genomes of this domain as a consequence of long evolutionary and diverged metabolic events. However, a close phylogenetic proximity between archaea and bacteria was observed for sharing metabolic functions by protein family divergence. A close phylogenetic boundary between methanogens and methylotrophs was revealed the reverse methanogenesis hypothesis. Energetic metabolism of methanogens are shared their ancestral behaviors with primitive proteobacteria. A mosaic nature of some functional domains of protein superfamilies was observed between mesophilic methanogens and thermophilic archaea/halophilic archaea, which may be resulted due to selection pressures. Overall results revealed that the phylogenetic analysis of protein superfamilies obtained from archaeal domain provided a standpoint for the evolutionary history of some key metabolic modules among methanogens.

* Tel: +91-431-2407071 Fax: +91-431-2407045 Email: pchellapandi@gmail.com.

Therefore, molecular evolutionary hypothesis from this finding would provide a new horizon for growth of research in advancement of metabolomics and metabolic engineering in methanogens.

In: Biochemistry Researcher Biographical Sketches ... ISBN: 978-1-62081-308-9
Editors: J. T. Walker and C. M. Jones

Chapter 86

Aptamers: The New Biorecognition Element for Proteomic Biosensing

Mònica Mir*
Nanobioengineering Laboratory, Institute for Bioengineering of Catalonia (IBEC), Barcelona Science Park, Centro de Investigación Biomédica en Red de Bioingeniería, Biomateriales y Nanomedicina, (CIBER-BBN), Spain

Research Summary

Aptamers are single stranded artificial nucleic acid ligands that can be generated against almost any kind of target, such as ions, metabolites aminoacids, drugs, toxins, proteins or whole cells. They are isolated from combinatorial libraries of synthetic nucleic acids by an iterative process of adsorption, recovery and amplification, know as SELEX (Systematic Evolution of Ligands by EXponential enrichment) process.

Aptamers, the nucleic acid equivalent to antibodies, are easy to synthesise, is not required the use of animals for its synthesis, for this reason it can be developed again toxins and small molecules that do not produce immune response in animals and can be tuned for affinity in closer to assay conditions permitting recognition out of the physiological state. So, aptamers posses numerous advantages that make them preferred candidates as biorecognition elements.

In view of the advantages and simple structure of aptamers, they have been used in a wide range of applications such as therapeutics, diagnosis, chromatography, environmental detection, among other.

* Corresponding author; Baldiri Reixac, 10-12, 08028 Barcelona, Spain, Phone; +34934037178, Fax; +34934037181. E-mail: mmir@ibec.pcb.ub.es.

In: Biochemistry Researcher Biographical Sketches ... ISBN: 978-1-62081-308-9
Editors: J. T. Walker and C. M. Jones

Chapter 87

Effect on Heavy Oils by Bacteria

Ruixia Hao, Guanyu Wang and Anhuai Lu
School of Earth and Space Sciences,
Peking University, Beijing, China

Research Summary

The objective of this chapter is to demonstrate the basic characteristics of halophilic strain TM-1 and Bacillus SP-5 which were isolated from a reservoir of the Shengli oil field in East China and evaluate the effects of strains TM-1 and SP-5 on different heavy oils. Strain TM-1 is a halophilic gram-positive coccus and strain SP-5 is gram-positive rod with spores. The cells of strains TM-1 and SP-5 were grown at temperatures up to 58°C in the neutral to alkaline pH range. They could grow steadily, and produce various metabolites, have various surface features and use different organic substrates (acetate, D-glucose, fructose, glycerol, maltose, pyruvate, starch, sucrose, and xylose). Laboratory studies have demonstrated that the strains TM-1 and SP-5 affected different heavy oils; converted and degraded various components and changed the physical and chemical properties of heavy oils. The bioconversion of heavy oils leads to an enrichment in lighter hydrocarbons and an overall redistribution of these hydrocarbons. The interactions of microorganisms with heavy oils are variable and depend on the microbial species and the chemical compositions of heavy oils.

In: Biochemistry Researcher Biographical Sketches ... ISBN: 978-1-62081-308-9
Editors: J. T. Walker and C. M. Jones

Chapter 88

Effect of Forest Environments on Human Urinary Adrenaline

Qing Li* and Tomoyuki Kawada
Department of Hygiene and Public Health, Nippon Medical School, Tokyo, Japan

Research Summary

Humans have enjoyed forest environments for ages because of the quiet atmosphere, beautiful scenery, mild climate, and fresh, clean air. Empirically, forest environments may reduce stress and have relaxing effect on human. Adrenaline is released from the adrenal medulla, and adrenaline levels increase under circumstances of novelty, anticipation, unpredictability, and general emotional arousal. Measurement of free adrenaline in urine provides a reliable measure of the circulating concentration of adrenaline in the bloodstream and thus is a measure of sympathoadrenal medulla activity. To explore the relaxing effect of forest environments on humans, we investigated the effect of trips to forest parks on urinary adrenaline and noradrenaline in both male and female subjects. We found that three-day/two-night trips to forest parks significantly reduce the concentration of urinary adrenaline and/or noradrenaline. Moreover, a day trip to a forest park also has a similar effect on urinary adrenaline. The Profile of Mood States (POMS) test showed that forest environments significantly increased the score for vigor and decreased the scores for anxiety, depression, anger, confusion, and fatigue suggesting that the subjects were under conditions of lower stress during the forest trips, which support the findings on urinary adrenaline. Other studies have reported that forest environments reduce the concentration of cortisol in saliva, reduce prefrontal cerebral activity, reduce blood pressure, and stabilize autonomic nervous activity in humans, which also support our findings on urinary adrenaline. These findings indicate that forest environments may reduce stress and have relaxing effect on human by reducing adrenaline level.

* Corresponding author: Qing Li, MD, Ph.D, Department of Hygiene and Public Health, Nippon Medical School, 1-1-5 Sendagi, Bunkyo-ku, Tokyo 113-8602, Japan. Tel: ++81-3-3822-2131, Fax: ++81-3-5685-3065. e-mail: qing-li@nms.ac.jp. URL: http://forest-medicine.

In: Biochemistry Researcher Biographical Sketches ... ISBN: 978-1-62081-308-9
Editors: J. T. Walker and C. M. Jones

Chapter 89

Microarray-Based Assay for Screening Kinase Inhibitors by Biotinylated Gold Nanoparticle Probes

Tao Li and Zhenxin Wang*
State Key Laboratory of Electroanalytical Chemistry, Changchun Institute of Applied Chemistry, Chinese Academy of Sciences, Changchun, China
Graduate School of the Chinese Academy of Sciences, Beijing, China

Research Summary

Here, a microarray-based spectroscopic assay with three readout principles, fluorescence, surface resonance Raman (SERS) and resonance light scattering (RLS), for studying kinase functionality and screening kinase inhibitors has been reported. In this assay, the phosphorylation and inhibition events are marked by biotinylated anti-phosphoserinen antibodies, and gold nanoparticles are attached to the antibodies by standard streptavidin-biotin chemistry, followed by silver deposition for RLS and SERS signal enhancement. The streptavidin conjugated fluorescein is used as fluorescent probe. The utility of this assay to high-throughput screening was demonstrated with the interactions of the α-catalytic subunit of cyclic adenosine 5′-monophosphate (cAMP) dependent protein kinase (PKA) with a commercial inhibitor library, a collection of 80 potential kinase inhibitors, and satisfactory results were obtained. In addition, quantitative determination of binding strength and the inhibiting type (type I) of these inhibitors are also demonstrated by the adenosine 5′-triphosphate (ATP) competing assays.

* Corresponding author. E-mail: wangzx@ciac.jl.cn; Fax:(+86) 431-85262243.

In: Biochemistry Researcher Biographical Sketches ...
Editors: J. T. Walker and C. M. Jones

ISBN: 978-1-62081-308-9

Chapter 90

ATP at a Crossroads: Cell Life or Death?

A. Atlante, A. Bobba, E. Marra and S. Passarella
[1]Istituto di Biomembrane e Bioenergetica, CNR, Bari, Italy
[2]Dipartimento di Scienze per la Salute, Università del Molise, Campobasso – Italy

Research Summary

In the good old days, the picture of the mitochondrial universe was simple. Good mitochondria produce ATP, which is then used for many cell functions as well as to synthesize the multi-thousands of types of macromolecules that are necessary for the cell to exist.

These good old days of simple biology are long gone in the wake of new findings: the ATP level or the available energy has been identified as one of the key factors determining the cell death via apoptosis or necrosis. The finding that depletion of ATP during apoptosis prevents the execution of apoptosis, but converts the cell death pathways to necrosis, is consistent with the fact that the apoptotic program has energy-requiring steps, including: (i) apoptosome complex formation and processing of pro-caspase-9; (ii) chromatin condensation and apoptotic body formation; (iii) phosphorylation of kinases. Thus, in apoptosis the role of both ATP, mostly produced in the mitochondria by oxidative phosphorylation, and the mitochondrial adenine-nucleotide translocator, through which the ATP reaches the cytosol, is crucial.

Here we will discuss: (i) how ATP level decides the mode of cell death; (ii) the steps which require ATP in apoptosis; (iii) the role of adenine-nucleotide translocator which acts driving the ATP export from mitochondria, as a component of the mitochondrial permeability transition pore, or as the specific target for the toxic fragment of tau protein whose production occurs in the cell death process.

Primary rat cerebellar granule cell culture will be considered as a model cell system and will be investigated en route to apoptosis. In order to favour the understanding of the general issue, other cellular and animal systems will also be mentioned.

In: Biochemistry Researcher Biographical Sketches ... ISBN: 978-1-62081-308-9
Editors: J. T. Walker and C. M. Jones

Chapter 91

Bioenergetics of Closed Ecological Systems: Effects of Carbon Sources

Frieda B. Taub
School of Aquatic and Fishery Sciences, University of Washington, Seattle, Washington, US

Research Summary

The goal of these studies was to estimate O_2 and CO_2 changes during light:dark cycles in aquatic Closed Ecological Systems of algae, grazers (*Daphnia*), and microbes and, using indirect calorimetry, to calculate the daily energy exchanges of the entire ecosystem. Closed Ecological Systems (CESs) are open to energy exchanges (light and heat), but closed to material exchanges (sealed from the earth's atmosphere). Here, I report on the results of using inorganic or organic carbon sources ($NaHCO_3$, cellulose, or no-added-carbon) on ecosystems consisting of three species of algae (*Ankistrodesmus*, *Scenedesmus*, and *Selenastrum*) with or without grazers (*Daphnia magna*) and associated microbes in chemically defined media of different concentrations, at 20°C on light:dark cycles of 12 hours each. Some of the CESs had O_2 and pH recorded every five minutes, and biological attributes of in vivo fluorescence (as a measure of algal abundance) and *Daphnia* numbers were counted weekly in both instrumented and non-instrumented CESs.

With $NaHCO_3$ as the carbon source, the algal growth occurred quickly, and grazers reproduced quickly (peaked at ~ 7-14 days), overgrazed their food supply, and eventually died out; algae subsequently increased. Cellulose provided slower growing algae and grazer populations (peaked ~ 28–35 days), but the grazer populations were of greater magnitude and persisted longer. The cellulose probably was metabolized and provided CO_2 for algal growth, which in turn fed the *Daphnia*. Without additional carbon ($NaHCO_3$ or cellulose), algae and *Daphnia* were of reduced abundance.

The O_2 and TCO_2 (Total Dissolved Inorganic Carbon, estimated from alkalinity and pH) did not always agree in magnitude or pattern. In the absence of grazers, pH tended to increase each lighted period more than it decreased each dark period, so that pH became > 10; CO_2

and HCO_3^- were largely depleted, and CO_3^{--} became the major carbon ion. In the presence of active grazers, pH was reduced to 6–7, CO_2 reappeared, and CO_3^{--} was reduced. The algae and the *Daphnia* demonstrated classical "predator-prey" dynamics. When grazers decreased, algal growth occurred and pH shifted to be more alkaline. O_2 behaved initially in a similar mode, increasing more each lighted period than decreasing each dark period (as algal biomass was accumulating), but later tended to have equal increases and decreases each period, so it was relatively stable. However, there are still technical issues to be resolved before O_2 and CO_2 can be equated with energy exchanges. Suggestions for creating better CESs are provided.

In: Biochemistry Researcher Biographical Sketches ... ISBN: 978-1-62081-308-9
Editors: J. T. Walker and C. M. Jones

Chapter 92

The Role of Mitochondria in the Glucose Metabolism

***L. de Bari*[1], *A. Atlante*[1] *and S. Passarella*[2]**
[1]Istituto di Biomembrane e *Bioenergetica*, CNR, Bari, Italy
[2] Dipartimento di Scienze per la Salute, Università del Molise, Campobasso, Italy

Research Summary

Although mitochondria have been the object of intensive study over many decades, some aspects of their metabolism remain to be fully elucidated, including the role of mitochondria in glucose metabolism.

In this regard in addition to glycolysis and gluconeogenesis, the methylglyoxal pathway must be taken in consideration. Thus we will review here the novel insights arisen from some recent achievements on the role of mitochondria on glucose metabolism.

In particular we will deal with:

1) The mitochondrial shuttles devoted to transfer reducing equivalents from the cytoplasm to mitochondria including the glycerolphosphate/dihydroxyacetone-phosphate, malate/aspartate, and more importantly, the malate/oxaloacetate and L-lactate/pyruvate shuttles
2) Transport and metabolism in mitochondria of phosphoenolpyruvate with a special emphasis to the mitochondrial pyruvate kinase recently reported.
3) Transport and metabolism in mitochondria of L-lactate with a special emphasis to the mechanism of L-lactate dependent gluconeogenesis due to both the mitochondrial L-lactate dehydrogenase and several L-lactate translocators.
4) Transport and metabolism in mitochondria of D-lactate with emphasis to the D-lactate translocators and the D-lactate dehydrogenase which could account for the removal of the toxic methylglyoxal from cytosol, as well as for D-lactate-dependent gluconeogenesis.

All together the reported findings will allow for a new scenario for the role of mitochondria in glucose metabolism.

In: Biochemistry Researcher Biographical Sketches … ISBN: 978-1-62081-308-9
Editors: J. T. Walker and C. M. Jones

Chapter 93

Mitochondrial Dysfunction Produced by Zn (II) or Selenite: A Comparison with Cd (II) and Ca (II)

Elena A. Belyaeva*
Laboratory of Comparative Biochemistry of Inorganic Ions, Sechenov Institute of Evolutionary Physiology and Biochemistry of Russian Academy of Sciences, St. Petersburg, Russia

Research Summary

To better understand the molecular mechanism(s) of mitochondrial dysfunction produced by heavy metals, we compared the action of Zn with effects of Cd, Ca, or Se (added as Na_2SeO_3) on isolated rat liver mitochondria used as a model system. With the help of O_2, TPP^+, K^+ and Ca^{2+}-selective electrodes, the simultaneous monitoring of four bioenergetics parameters: respiration, mitochondrial transmembrane potential ($\Delta\Psi_{mito}$), K^+ and Ca^{2+} fluxes was conducted in the presence of the metals/metalloids under test and different mitochondrial effectors to underscore the cause/consequence relationships underlying the mitochondrial dysfunction. Changes in redox state of pyridine nucleotides, matrix swelling, K^+ and Ca^{2+} release and the $\Delta\Psi_{mito}$ loss produced by Zn or Cd were significantly inhibited by cyclosporine A (CsA), i.e. a potent pharmacological inhibitor of mitochondrial permeability transition (MPT) pore, but even in its presence a slow dissipation of the membrane potential took place. In contrast to this, the $\Delta\Psi_{mito}$ loss observed in the presence of Se or Ca was completely eliminated by CsA. At the same time, the strong sustained stimulation of the mitochondrial basal respiration induced by Se only partially decreased by this MPT pore inhibitor. The conducted experiments showed also that stigmatellin (i.e. a selective inhibitor of complex III of mitochondrial electron transport chain, ETC) exhibited the opposite effects on the mitochondrial swelling induced by the metals/metalloids in the isotonic sucrose medium in the presence of ascorbate and TMPD (ETC complex IV substrates). In particular, stigmatellin

* Corresponding author: Fax: +7-812-552-3012; E-mail: *alenab61@mail.ru*; belyaeva@EB10330.spb.edu.

depressed partially the matrix swelling, produced by Zn, Cd or Ca, but sharply enhanced the swelling evoked by Se. The protective action of the MPT and ETC effectors against the mentioned above mitochondrial dysfunction was overridden by increase of the metals/metalloids load. As a result, similarities and differences in the toxic effects of the metals/metalloids under test as well as sequences of events manifested in the mitochondrial dysfunction were analyzed.

In: Biochemistry Researcher Biographical Sketches ...
Editors: J. T. Walker and C. M. Jones

ISBN: 978-1-62081-308-9

Chapter 94

Extramitochondrial Aerobic Metabolism in Retinal Rod Outer Segment Disks

Isabella Panfoli[*]
Department of Biology-DIBIO, University of Genoa, Genova, Italy

Research Summary

Vertebrate retinal rod Outer Segments (OS) are the site of visual transduction, an energy demanding process, for which energy supply is still puzzling. In fact glycolysis or diffusion of ATP from the mitochondria of the inner segment do not seem adequate to supply phototransduction. We have previously reported data suggestive of the presence of an aerobic metabolism in rod OS, consistently with the results of a proteomic study, where the five redox chain complexes where identified in disks. The presence of mitochondrial Respiratory Chain components in extra-mitochondrial membranes is not new but the function of these proteins has not been fully clarified. By contrast, our previous results on bovine rod outer segments and brain myelin vesicles, that are devoid of mitochondria, cogently suggested that these express functional mitochondrial redox apparatus.

Here we show that rod outer segment disks consume O2 in the presence of diverse respiring substrates, and contain the whole biochemical machinery for the complete combustion of glucose. Results are discussed in terms of quantitative bioenergetics and the implications of an exportability of the mitochondrial machinery to nervous system sites where ATP need is higher are also considered. Data shed light on many retinal pathologies related to energy supply in rod OS.

[*] Corresponding author: University of Genova, Faculty of Sciences, Biology Department (www.dibio.unige.it), V.le Benedetto XV, 3, 16132 GENOVA, Phone + 39 010 353 7397, Fax + 39 010 353 8153, E-mail: Isabella.Panfoli@unige.it.

In: Biochemistry Researcher Biographical Sketches ...
Editors: J. T. Walker and C. M. Jones
ISBN: 978-1-62081-308-9

Chapter 95

Bioenergetics and Male Infertility: From Basic Science to Clinical Andrology

***Alexandra Amaral*[1], *Ana Paula Sousa*[1,2], *Pedro Caballero Campo*[3], *Pedro Caballero Peregrín*[3] *and João Ramalho-Santos*[*1]**

[1]Center for Neuroscience and Cell Biology, Department of Life Sciences, University of Coimbra, Coimbra, Portugal
[2]Human Reproduction Service, University Hospitals of Coimbra, Portugal
[3]Human Reproduction Unit, Tambre Clinic and Tambre Foundation, Madrid, Spain

Research Summary

Human sperm is a terminally differentiated cell containing highly condensed DNA, and a paucity of cytoplasm and intracellular organelles. The single goal of a given spermatozoon is to achieve fertilization. Sperm bioenergetics are crucial for fertilization potential, given that ATP is required for a myriad of functions, from movement to signaling processes involved in capacitation and the acrosome reaction. In fact, several bioenergetics-related parameters have been proposed to aid in the evaluation of sperm functionality and in the diagnosis of male infertility. Interestingly, the nature of the ATP that fuels distinct sperm–specific processes has been hotly debated for decades. The discussion usually involves the relative importance of two sugar metabolic pathways: glycolysis (restricted to the fibrous sheath, in the sperm tail) and oxidative phosphorylation (OXPHOS, occurring in mitochondria, localized in the sperm midpiece). Although a growing body of data seems to suggest that glycolysis has a preponderant role in human sperm, OXPHOS seems to also be relevant, and other metabolic pathways are most likely involved. In addition, oxidation reactions occurring in sperm mitochondria can be accompanied by the production of reactive oxygen species (ROS), which, although important for normal signaling pathways, may also cause sperm dysfunction.

* Corresponding author: jramalho@ci.uc.pt, Center for Neuroscience and Cell Biology, Department of Life Sciences, University of Coimbra, P.O. Box 3046, 3001-401 Coimbra, Portugal.

The importance of human sperm mitochondria will be discussed, together with the clinical relevance of monitoring a number of mitochondrial features, from the genome and proteome levels (sperm mitochondrial DNA and proteins) to cellular features (mitochondrial membrane potential).

In: Biochemistry Researcher Biographical Sketches ... ISBN: 978-1-62081-308-9
Editors: J. T. Walker and C. M. Jones

Chapter 96

Identification of Specific Mitochondrial Proteins Forming Stable Adducts with 4-Hydroxynonenal within Cardiac Tissue of Type-I Diabetic Animals: Implications for Bioenergetics Dysfunction and Onset of Diabetic Cardiomyopathy[#]

Gregg DiNuoscio[1], Chao Yuan[2], Ossama Lashin[1] and Andrea Romani[1]*

[1]Department of Physiology and Biophysics, Case Western Reserve University,
[2] Case Proteomics Center Case Western Reserve University–Cleveland, OH, US

Research Summary

Diabetes is a metabolic syndrome associated with the onset of numerous complications. Among these complications, diabetic cardiomyopathy represents one of the most common causes of death in diabetic patients. In recent years, experimental evidence has indicated a major role of reactive oxygen species (ROS) and lipid peroxidation products in the ethio-pathogenesis of diabetes and its complications. Disagreement, however, exists about the sites and mechanisms involved in ROS formation, the role of hyperglycemia in exacerbating their formation, and the effectiveness of antioxidant levels in counteracting the formation of ROS and peroxidation products.

[#] A version of this book was also published as a chapter in *Cell Respiration and Cell Survival: Processes, Types and Effects* published by Nova Science Publishers, Inc. It was submitted for appropriate modifications in an effort to encourage wider dissemination of research.

[*] Corresponding author: Phone: 1-216-3681625 Fax: 1-216-3685586 e-mail: amr5@po.cwru.edu

Studies from our laboratory have provided compelling evidence that type-I diabetes is associated with the endogenous production of 4-hydroxynonenal (HNE). Generated via lipid peroxidation, this highly reactive aldehyde rapidly reacts and forms stable adducts with specific proteins within the cardiac mitochondria of diabetic rats. One of the modified proteins has been successfully identified as the FAD containing subunit of succinate dehydrogenase, and the defect has been correlated to a comparable decline in succinate-supported oxygen consumption in the intact mitochondria and in complex II activity in the purified complex. The process leading to HNE production appears to be independent of glycemia level, but is strictly associated to the decline in circulating insulin level. Streptozotocin-injected animals presenting levels of circulating insulin higher than 65 pmol/L show low levels of HNE-induced modification of mitochondrial proteins and near normal oxygen consumption in intact mitochondria despite presenting hyperglycemia levels comparable to those of frankly diabetic animals (i.e. >400 mg glucose/dl). In contrast, frankly diabetic animals present endogenous circulating insulin levels below 25 pmol/L, high levels of HNE-induced modifications, and defects in mitochondrial respiration. Supplementation of the latter group of animals with exogenous insulin for a minimum of 2 weeks results in a marked decrease in HNE-mediated adducts and the restoration of succinate dehydrogenase activity to levels comparable to those observed in non-diabetic animals. The identification of two distinct pools of hyperglycemic animals that present or lack HNE-induced modifications and related mitochondrial dysfunction has provided a new rationale to explain some of the functional inconsistencies observed in mitochondria purified from various tissues of diabetic animals.

Utilizing a proteomics approach, we have identified about 12 mitochondrial proteins that form stable adducts with HNE. Determining to what extent these proteins are functionally impaired by the formation of adducts with HNE will undoubtedly help to shed light on the role of mitochondrial impairment in the onset of diabetic cardiomyopathy.

In: Biochemistry Researcher Biographical Sketches ... ISBN: 978-1-62081-308-9
Editors: J. T. Walker and C. M. Jones

Chapter 97

Simultaneous Fluorescence and Reflection Confocal Microscopy Study of Living Osteoblast Bioenergetics as a Tool for the Design of Surface Topography of Dental Implants[#]

Mercedes Salido*[*1], J. Ignacio Vilches-Perez[2] and Jose Vilches[1]
[1]Cell Culture and Confocal Microscopy Unit, Department of Histology, School of Medicine, University of Cadiz, Spain
[2]Department of Oral Surgery, School of Dentistry, University of Seville, Spain

Research Summary

Confocal laser scanning microscope (CLSM) appears to be a valuable and non invasive tool for simultaneous 3D visualization of both cell morphology and structure and material scaffolds. The CLSM is also capable of creating excellent images using backscattered or reflected light. Confocal backscattered imaging is particularly useful for creating 3D images of the surface of materials and profilometric studies. For the quantification of maximal and minimal values in surface profile, distance between peaks in each surface, width of grooves and valleys, and axial ratio of the cells growing on the different surfaces were assessed. In addition to image data of cell morphology, microstructural data of scaffolds and cell-scaffold interaction can be readily obtained using this method. The quality of cell adhesion to the dental implant surface determines the tissue integration, and the surface can directly influence osteoblast adherence, attachment, spreading and metabolism, modifying and controlling the

[#] A version of this book was also published as a chapter in *Dental Materials Research* published by Nova Science Publishers, Inc. It was submitted for appropriate modifications in an effort to encourage wider dissemination of research.

[*] Corresponding author. Cell Culture and Confocal Microscopy Unit., c/Dr Marañon 3, 11002 Cadiz. Spain. Tel: +34 956015832; fax: +34 956015265. E-mail address: mercedes.salido@uca.es (M Salido)

osseointegration process. The combinated use of reflectance and fluorescence modes is of capital interest to provide detailed images of living cell interactions with underlying surfaces in biomedical devices as dental implants, building the cellular and molecular basis of mechanotransduction, a process where high amounts of energy are demanded. Consistently with this, mitochondria that accumulate at sites of development of focal adhesions, mechanosensory devices, exhibit higher membrane potential compared with mitochondria that are not at those sites.

The methodology described herein is useful to directly visualize and quantify the role of underlying environmental cues for force-generating and anchoraging activities on mitochondrial bioenergetics in human living osteoblasts, growing on a number of customized biomaterials surfaces, and permit the use as a tool for designing the surface quality in order to provide the ideal surface for osseointegration of dental implants.

In: Biochemistry Researcher Biographical Sketches … ISBN: 978-1-62081-308-9
Editors: J. T. Walker and C. M. Jones

Chapter 98

Screening and Studying Photosynthetic Mutants: Basics and Beyond

Xenie Johnson**, *Laura Houille-Vernes and Jean Alric
UMR 7141, CNRS et Université Pierre et Marie Curie (Paris VI),
Institut de Biologie Physico-Chimique, Paris, France

Research Summary

Among bioenergetics chains the photosynthetic process has been extensively studied using its very substrate, light, as a way to probe photosynthetic activity and to screen mutants. The green algae *Chlamydomonas reinhardtii* was one of the first model organisms used for generating non-photosynthetic mutants. Chlamydomonas is a facultative phototroph, as a consequence, the mutants missing a major component of the photosynthetic chain grow equally well on a medium supplemented with a reduced carbon source and remain green even in the dark. These mutants have been useful not only in dissecting photosynthetic electron transfer but also for identifying the various subunits of the protein complexes and understanding chloroplast biogenesis (the concerted accumulation of these subunits). However, alternative pathways of electron transfer such as chlororespiration are less well characterised in Chlamydomonas because of the lack of mutants. With the nuclear genome sequence recently available, it is timely to investigate the function of the large number of unknown proteins that may putatively interact with photosynthetic processes. Here we discuss the rationale behind the selection of mutants affected in new pathways.

* Corresponding author: jean.alric@ibpc.fr,

In: Biochemistry Researcher Biographical Sketches … ISBN: 978-1-62081-308-9
Editors: J. T. Walker and C. M. Jones

Chapter 99

Bioengineering Approaches for Aetiological Research in Severe Pressure Ulcers

Amit Gefen and Eran Linder-Ganz
Department of Biomedical Engineering,
Faculty of Engineering, Tel Aviv University, Israel

Research Summary

Pressure ulcers (PU) are soft tissue injuries associated with excessive tissue deformation, ischemia, impaired metabolic activity and insufficient lymph drainage, which are all caused by prolonged and intensive mechanical loads. Also called bed-sores or decubitus ulcers, PU range in severity from superficial skin irritation and lesions in subcutaneous and fat tissues to deep muscle necrosis. Serious PU involving extensive sub-dermal tissue damage, where damage originates in muscle tissue, were recently termed “deep tissue injury" (DTI) by the US National Pressure Ulcer Advisory Panel (NPUAP), and are attracting growing attention by the medical community.

In: Biochemistry Researcher Biographical Sketches ... ISBN: 978-1-62081-308-9
Editors: J. T. Walker and C. M. Jones

Chapter 100

Ion Beam Bioengineering Research in Thailand

S. Anuntalabhochai[1,*], L. D. Yu[2,3,*,#] and T. Vilaithong[3]

[1]Molecular Biology Laboratory, Department of Biology, Faculty of Science, Chiang Mai University, Thailand

[2]Plasma and Beam Physics Research Facility, Department of Physics and Materials Science, Faculty of Science, Chiang Mai University, Thailand

[3]Thailand Center of Excellence in Physics, Commission on Higher Education, Thailand

Research Summary

This review summarizes recent progress of a newly developed novel bioengineering technology, namely low-energy ion beam bioengineering (IBB), achieved at Chiang Mai University, Thailand. Low-energy IBB has demonstrated powerful impacts and highly potential applications on biology, agriculture, horticulture and life science owing to multiple-factor interaction between energetic ions and biological organisms as well as low-cost and convenience in operation. Since late 1990s, IBB research programs have been vigorously carried out at Chiang Mai University. A group of scientists consisting of physicists, biologists, horticulturists, agriculturists, chemists, and medical scientists have developed special IBB ion beam lines and relevant techniques to use low-energy ion beam bombardment of biological organisms to induce mutation breeding and gene transfer. Besides the IBB applications, research interests are also focused onto basic mechanisms on ion interaction with DNA and biological cells to reveal physics and biology involved in the ion beam inductions of mutation and gene transfer. The results have been applied to serve developments of local agriculture and horticulture and promote national scientific research qualities. The contents include introductions to low-energy IBB techniques and facilities, and chapters on ion beam induced gene transfer, ion interaction with the cell envelope, ion beam induced mutation, and ion interaction with living cells and DNA.

---

* Both authors contributed equally.

Corresponding author: Tel.: +66 53 943379, Fax: +66 53 222776, Email: yuld@fnrf.science.cmu.ac.th.

In: Biochemistry Researcher Biographical Sketches ... ISBN: 978-1-62081-308-9
Editors: J. T. Walker and C. M. Jones

Chapter 101

Analysis of Shell Structures Applying Triangular Finite Elements

C. W. Solomon
Department of Mechanical Engineering,
University of Nebraska, Lincoln, Nebraska, US

Research Summary

Over the years many shell finite elements has been developed for applications in aerospace, automotive and shipbuilding industries. Misuse or abuse of some of these shell finite elements in the relatively mature shell finite element technology is not uncommon. Some of these shell finite elements were based on the principles of classical shell theory in which the simplest one is the theory of Love. Others were based on intuitive or heuristic arguments. For reasons of economy, mathematical simplicity and accuracy, lower order flat triangular shell finite elements are popular among designers. This article is concerned with the review, development and application of triangular shell finite elements. Emphasis is on mixed formulation based lower order flat triangular shell finite elements. Finite element representation of shell structures is introduced. Linear analysis of static and dynamic, and nonlinear analysis of static and dynamic problems, are included in this article.

In: Biochemistry Researcher Biographical Sketches ... ISBN: 978-1-62081-308-9
Editors: J. T. Walker and C. M. Jones

Chapter 102

Bioengineering of Human Fetal Tissues for Clinical Use

Lee Ann Applegate[1,2*], *Nathalie Hirt-Burri*[1], *Corinne Scaletta*[1,2], *Jean-François Bauen*[3] *and Dominique P. Pioletti*[2]
[1]Cellular Therapy Unit, Department of Musculoskeletal Medicine (DALO), University Hospital of Lausanne (CHUV-UNIL), Switzerland
[2]Laboratory of Biomechanical Orthopedics, Center for Translational Biomechanics EPFL-CHUV-DAL, Ecole Polytechnique Fédérale de Lausanne, Switzerland
[3]Gynecology Department, Yverdon-les-Bains Hospital, Switzerland

Research Summary

Cultured primary fetal cells from one organ donation could possibly meet the exigent and stringent technical aspects for development of therapeutic products. These cell types have fewer technological limitations for cellular proliferation capacity (simple culture conditions) and maintenance of differentiated phenotype, and they also have low probability for transmission of communicable diseases. Master and Working Cell Banks (MCB, WCB) can be obtained from one fetal organ donation, permitting multiple tissues (skin, bone, cartilage, muscle and intervertebral disc) to be processed in short periods of time with identical methods to assure a stringent tracing of the processes for the production of standardized therapeutic agents. Clinical use of biologics from embryo and fetal tissues is relatively new and current legislation and ethics have some differences between countries to date. In addition, specific cell delivery systems for each tissue type can be adapted to the clinical application. Since it is the intention that banked primary fetal cells enhance the prospective treatment of hundreds of thousands of patients with only one organ donation, it is imperative to show consistency, traceability and safety of the processes including donor tissue selection, cell banking, cell testing and growth of cells in out-scaling for the preparation of bio-engineered products for clinical application.

[*] Correspondence:
Dr. Lee Ann Laurent-Applegate, Cellular Therapy Unit, Department of Musculoskeletal Medicine (DALO), University Hospital, CHUV, PAV 03, 1011 Lausanne, Switzerland, Tel. ++41 21 314 35 10, Fax ++41 21 887 84 14, Email: Lee.Laurent-Applegate@chuv.ch.

In: Biochemistry Researcher Biographical Sketches ...
Editors: J. T. Walker and C. M. Jones
ISBN: 978-1-62081-308-9

Chapter 103

Gene Cluster or Operon Design by Ordered Gene Assembly in *Bacillus Subtilis* (OGAB) Method

***Kenji Tsuge*[1*] *and Mitsuhiro Itaya*[2]**
[1]Assistant Professor, Institute for Advanced Biosciences, Keio University
[2]Professor, Institute for Advanced Biosciences, Keio University, Japan

Research Summary

In the post genome sequence era, metabolic engineering conducted by a number of genes is attracting great attention for not only basic research but also for application in production of beneficial materials for our lives. To carry this out, methods to handle a number of genes at a time are a prerequisite. However, the poor ability of *Escherichia coli*-based genetic methods to manipulate the reconstruction makes the steps time-consuming and laborious. *Bacillus subtilis*, a gram-positive soil bacterium, develops natural competence in certain physiological conditions. In the competence, this bacterium can take up exogenous DNAs, and can recombine these DNAs with the genome if there is a homologous sequence. By using this trait, a novel method to carry out gene assembly quickly and with greatest fidelity using *B. subtilis* was developed. The method, named Ordered Gene Assembly in *B. subtilis* (OGAB), facilitates the connection of more than 10 DNA fragments in plasmid in one-step using an intrinsic *B. subtilis* plasmid transformation system. In this chapter, the principle of and actual protocol for the OGAB method are written. Moreover, two types of gene assembly in terms of metabolic engineering are also described.

* To whom correspondence should be addressed: Kenji Tsuge, Institute for Advanced Biosciences, Keio University, 403-1 Nihonkoku, Daihoji, Tsuruoka, Yamagata, 997-0017, Japan, TEL +81-235-29-0526, FAX +81-235-29-0529, E-mail: ktsuge@ttck.keio.ac.jp.

In: Biochemistry Researcher Biographical Sketches … ISBN: 978-1-62081-308-9
Editors: J. T. Walker and C. M. Jones

Chapter 104

Bioengineering of Glass-Ceramics and Ceramics for Dental Restoration

W. Höland and V. Rheinberger
Ivoclar Vivadent AG, Bendererstr. 2, Li-9494 Schaan, Principality of Liechtenstein

Research Summary

The challenge of bioengineering glass-ceramics and ceramics for the fabrication of dental restorations, such as inlays, onlays, veneers, dental crowns and dental bridges, is to fulfill the needs of patients. The methods and principles of developing metal-based and metal-free restorations according to bioengineering principles will be demonstrated and clinical applications will be shown. Current efforts are concentrated on developing materials for metal-free restorations. Bioengineering principles are employed with the objective of developing materials with properties that are close to those of natural teeth. Therefore, the translucency, color, mechanical properties and chemical durability of these biomaterials have to be close to or even better than the properties of natural teeth. The authors will report on methods for developing materials that show a combination of different properties. Solid state reactions in various chemical systems which are prepared according to different methods will be used to fulfill the demands. Glass-ceramics and ceramics demonstrating high strength and toughness but also translucency and tooth color will be presented. The preferred and modern technologies of molding and machining will be shown.

In: Biochemistry Researcher Biographical Sketches … ISBN: 978-1-62081-308-9
Editors: J. T. Walker and C. M. Jones

Chapter 105

Development of a Diagnostic System for Osteoarthritis Using a Photoacoustic Measurement Method and Time-Resolved Auto-Fluorescence

Masato Sato [1*]***, Miya Ishihara*** [1,2]***, Genya Mitani***[1]***, Toshiharu Kutsuna***[1]***, Jeong Ik Lee***[1]***, Makoto Kikuchi***[2] ***and Joji Mochida***[1]

[1]Department of Orthopaedic Surgery, Surgical Science, Tokai University School of Medicine, Isehara, Kanagawa, Japan
[2]Department of Medical Engineering, National Defense Medical College, Saitama, Japan

Research Summary

We have developed new methods to measure some essential properties of cartilage: a photoacoustic measurement method and time-resolved fluorescence spectroscopy. These can be used to evaluate the outcomes of tissue-engineered cartilage, regenerated articular cartilage tissue after surgery, and the degenerated cartilage of osteoarthritis patients. A nanosecond-pulsed laser, which is completely noninvasive, is focused onto the target cartilage and induces a photoacoustic wave, which will propagate with attenuation, which is affected by the viscoelasticity of the surrounding cartilage. The decay time during which the amplitude of the photoacoustic wave is reduced by a factor of 1/e is the key numerical value used to characterize and evaluate the viscoelasticity and rheological behavior of the cartilage. In this study, we also investigated whether pulsed laser irradiation and the measurement of excited autofluorescence allow us to noninvasively evaluate tissue characters in real time. Our

[*] Corresponding author:
Masato Sato, MD., PhD., Associate Professor and Research Director, Department of Orthopaedic Surgery, Surgical Science, Tokai University School of Medicine 143 Shimokasuya, Isehara, Kanagawa, 259-1193 Japan. Email: sato-m@is.icc.u-tokai.ac.jp, TEL: +81-463-93-1121 (ext 2320), FAX: + 81-463-96-4404

findings show that time-resolved laser-induced fluorescence spectroscopy is useful for evaluating tissue-engineered cartilage. This measurement system, predicated on the interactions between optics and living organs, is a suitable methodology for diagnosis during arthroscopy, because it allows the quantitative and multidirectional evaluation of the original function of the cartilage based on a variety of parameters.

In: Biochemistry Researcher Biographical Sketches ... ISBN: 978-1-62081-308-9
Editors: J. T. Walker and C. M. Jones

Chapter 106

Biotechnology and Agriculture

Jorge Fernandez-Cornejo
U.S. Department of Agriculture (USDA) - Economic Research Service (ERS)
Washington DC, US

Research Summary

Farmers adopting first-generation genetically engineered (GE) crops derive tangible benefits, even though not all benefits are reflected in standard measures of net returns. The impacts of GE crops vary with annual pest infestations, seed premiums, prices of alternative pest control programs, and any premiums paid for segregated (i.e., non-GE) crops.

In: Biochemistry Researcher Biographical Sketches ... ISBN: 978-1-62081-308-9
Editors: J. T. Walker and C. M. Jones

Chapter 107

Calcium Orthophosphate-Based Biocomposites and Hybrid Biomaterials

Sergey V. Dorozhkin*
Moscow, Russia

Research Summary

In this book, the state-of-the-art of calcium orthophosphate-based biocomposites and hybrid biomaterials suitable for biomedical applications is presented. This subject belongs to a rapidly expanding area of science and research because these types of biomaterials offer many significant and exciting possibilities for hard tissue regeneration. Through the successful combinations of the desired properties of matrix materials with those of fillers (in such systems, calcium orthophosphates might play either role), innovative bone graft biomaterials can be designed. The book starts with an introduction to locate the reader. Further, general information on composites and hybrid materials, including a brief description of their major constituents are presented. Various types of calcium orthophosphate-based bone-analogue biocomposites and hybrid biomaterials those are either already in use or being investigated for various biomedical applications are then extensively discussed. Many different formulations in terms of the material constituents, fabrication technologies, structural and bioactive properties, as well as both in vitro and in vivo characteristics have been already proposed. Among the others, the nano-structurally controlled biocomposites, those with nano-sized calcium orthophosphates, biomimetically-fabricated formulations with collagen, chitin and/or gelatin, as well as various functionally graded structures seem to be the most promising candidates for clinical applications. The specific advantages of using calcium orthophosphate-based biocomposites and hybrid biomaterials in the selected applications are highlighted. As the way from a laboratory to a hospital is a long one and the prospective biomedical candidates have to meet many different necessities, the critical issues and scientific challenges that require further research and development, have been examined, as well.

* Kudrinskaja sq. 1-155, Moscow 123242, Russia; e-mail: sedorozhkin@yandex.ru.

In: Biochemistry Researcher Biographical Sketches … ISBN: 978-1-62081-308-9
Editors: J. T. Walker and C. M. Jones

Chapter 108

Calcium Orthophosphate Cements and Concretes

Sergey V. Dorozhkin*
Moscow, Russia

Research Summary

In early 1980-s, researchers discovered self-setting calcium orthophosphate cements, which are a bioactive and biodegradable grafting material in the form of a powder and a liquid. Both phases after mixing form a viscous paste that after being implanted sets and hardens within the body as either a non-stoichiometric calcium deficient hydroxyapatite (CDHA) or brushite, sometimes blended with unreacted particles and other phases. As both CDHA and brushite are remarkably biocompartible and bioresorbable (therefore, in vivo they can be replaced with a newly forming bone), calcium orthophosphate cements represent a good correction technique of non-weight-bearing bone fractures or defects and appear to be very promising materials for bone grafting applications. Besides, these cements possess an excellent osteoconductivity, molding capabilities and easy manipulation. Nearly perfect adaptation to the tissue surfaces in bone defects and a gradual bioresorption followed by new bone formation are additional distinctive advantages of calcium orthophosphate cements. Furthermore, reinforced cement formulations are available, which in a certain sense might be described as calcium orthophosphate concretes. The concepts established by calcium orthophosphate cement pioneers in the early 1980-s were used as a platform to initiate a new generation of bone substitute materials for commercialization. Since then, advances have been made in the composition, performance and manufacturing; several beneficial formulations have already been introduced as a result. Many other compositions are in experimental stages. In this review, an insight into calcium orthophosphate cements and concretes, as excellent biomaterials suitable for both dental and bone grafting application, has been provided.

* Corresponding author: E-mail: sedorozhkin@yandex.ru.

In: Biochemistry Researcher Biographical Sketches … ISBN: 978-1-62081-308-9
Editors: J. T. Walker and C. M. Jones

Chapter 109

Hydrogels in Biology and Medicine

J. Michalek, M. Pradny, K. Dusek, M. Duskova, R. Hobzova and J. Sirc
Institute of Macromolecular Chemistry AS CR, v.v.i., Prague, Czech Republic

Research Summary

The range of materials used for biomedical applications is very broad. This means that the demands on their properties are very diverse depending on various medical areas and applications. Moreover, it is often necessary to have available materials with the possibility to set the required parameters very precisely in very wide ranges. Because of the similar mechanical behaviour of hydrogels with that of living tissues and their good compatibility and ability of hydrogels to swell in water, the hydrogels are often used in biomedical applications.

Hydrogel polymers are natural or synthetic hydrophilic crosslinked polymers favourably interacting with water. The swelling degree (the water content in equilibrium-swollen gel) is a function of polymer hydrophilicity and the degree of crosslinking and has influence on a number of physical and chemical properties (e.g., refractive index, transport properties). In biomedical applications, the physical structure and morphology of the hydrogels is adjusted to the targeted performance and both homogenous and heterogeneous materials with porous, nanofiber, or nanoparticle structures are used.

One of the oldest and still widely used biomedical applications of hydrogels is the contact lens. Hydrogels are successfully applied as a variety of implants in surgery, ophthalmology, otorhinolaryngology, neurology, urology, gynaecology etc. In addition, hydrogels are used to cover wounds, burns, trophic defects, or as a two-dimensional supports for cultivation and potential transplantation of cells or as three-dimensional scaffolds for tissue engineering and cell therapy.

In some applications, particularly, in tissue engineering, the biodegradable materials (hydrogels) are used. The advantage is that after the fulfilment of their tasks (e.g., proliferation of cell culture), they disintegrate (hydrolytically or enzymatically) and, subsequently, they are eliminated from the organism. Controlled transport and release of drugs is a separate biomedical area.

Hydrogels for biomedical applications can be generally classified in different categories depending on their interaction with the living tissue. The hydrogels are used either in direct contact with tissues (implants, injection needle, infusion sets, bandages, suture materials, carrier of drugs, cell or tissue culture) or in indirect contact with tissues (orthesis, medical devices, air filters, sanitary supplies, etc.).

The matter of this chapter is arranged in the following subchapters:

1) Introduction
2) Structure of hydrogels in relation to the formation conditions
3) Swelling and mechanical properties of hydrogels
4) Contact lenses
5) Intraocular lenses
6) Functional implants
7) Embolization of blood vessels
8) Wound-dressing
9) Conductive supports for contact electrodes
10) Hydrogels for tissue engineering

In: Biochemistry Researcher Biographical Sketches ... ISBN: 978-1-62081-308-9
Editors: J. T. Walker and C. M. Jones

Chapter 110

Biomaterials in Dentistry and Medicine

Clare Crowley[1*], Tony Pembroke[2] and Colin Birkinshaw[1]
[1]Department of Materials Science and Engineering, University of Limerick, National Technological Park, Limerick, Ireland
[2]Department of Chemistry and Environmental Sciences, University of Limerick, National Technological Park, Limerick, Ireland

Research Summary

The widespread use of biomaterials in medicine and dentistry is a relatively new phenomenon dating back to the 1950's yet, today, an estimated 20 million individuals have an implanted medical device.

Despite the huge impact that biomaterials have had on patients' quality of life, improvements in device performance and the development of alternatives to augment available therapies are continuously being sought. Clinical demand, advances in molecular and cell biology and the increased understanding of the role of the tissue-material interface on clinical performance has led to a metamorphosis of the biomaterials' field over the past 25 years. This has resulted in a change in the nature of biomedical devices from being biologically passive to actively integrated.

This chapter explores the development and application of biomaterials over the past 25 years, examining the current clinical demand, the scientific rationale, and the technical challenges to be overcome. As biomaterials are applied in reconstructive surgery and tissue regenerative therapies, these areas are explored with specific examples of recent developments and current research activity used to illustrate the changing perspectives.

* Corresponding author: Department of Materials Science and Engineering, University of Limerick, National Technology Park, Limerick, IRELAND Email: Clare.Crowley@ul.ie, Tel: 00-353-61-2, Fax: 00-353-61-213529.

In: Biochemistry Researcher Biographical Sketches … ISBN: 978-1-62081-308-9
Editors: J. T. Walker and C. M. Jones

Chapter 111

Biomaterials in Blood-Contacting Devices: Complications and Solutions

***Meng-Jiy Wang*[1] *and Wei-Bor Tsai*[2]**
[1]Department of Chemical Engineering,
National Taiwan University of Science and Technology
[2]Department of Chemical Engineering, National Taiwan University

Research Summary

All blood-contacting medical devices in use today are subjected to some degree poorer blood compatibility than the native artery. Hemostatic mechanism, arresting bleeding from injured blood vessels, induces platelet adhesion and activation onto artificial biomaterials, which leads to undesirable outcomes such as blood clotting at the site of the implant, continual shedding of thrombi, and depletion of platelets from the blood stream. Such complications have hampered the clinical success of blood contacting devices, limiting the patency of small-diameter vascular grafts and making necessary the use of anticoagulants in patients undergoing extracorporeal bypass or synthetic heart valve implantation. Therefore, development of non-thrombogenic biomaterials is in great need for blood contacting devices. The current approaches mainly focus on surface modifications with biological anticoagulants such as heparin, or anti-fouling molecules like poly(ethylene oxide). In this review, we will first introduce the blood components involved in hemostasis and thrombosis, followed by the common biomaterials applied in blood-contacting devices. Next, the complications induced by the interactions between blood and biomaterials will be briefly addressed. Finally, the commonly used techniques for improving biomaterials' hemocompatibility will be expatiated.

In: Biochemistry Researcher Biographical Sketches … ISBN: 978-1-62081-308-9
Editors: J. T. Walker and C. M. Jones

Chapter 112

Biocompatibility of Dental and Medical Materials

Daniel Araki Ribeiro[1*], ***Ana Claudia Muniz Renno***[1] ***and Mariza Akemi Matsumoto***[2]
[1]Department of Biosciences, Federal University of Sao Paulo, UNIFESP, Santos, SP, Brazil
[2]Department of Health Sciences, School of Dentistry, Sacred Heart University, USC, Bauru, SP, Brazil

Research Summary

A number of biomaterials, organic or synthetic, are used in the various fields of medical or dentistry. Among them are the suture materials, osteosynthesis plates and screws, endosseous implants, bone substitutes, endodontic, restorative, prosthetic materials, and others. All of them are put in contact with biological tissues, deeply or superficially and most of them remain for lifetime. For this reason, all these materials should be carefully analyzed by means of biocompatibility tests, in order to evaluate materials' biological behavior. In this regard, the goal of this chapter is to present the biocompatibility of some biomaterials such as bioactive bone substitutes (ceramics, bioactive glasses and polymers) titanium endosseous implants, metal and polymeric osteosynthesis devices, restorative and endodontic materials based on our current research with the field.

* Corresponding author: Av. Ana Costa, 95, Vila Mathias, Santos – SP, Brazil, Zip code: 11060-001, Fone: +55 13 32218058, Fax: +55 13 32232592, e-mails: daribeiro@unifesp.br, daribeiro@pesquisador.cnpq.br.

In: Biochemistry Researcher Biographical Sketches ... ISBN: 978-1-62081-308-9
Editors: J. T. Walker and C. M. Jones

Chapter 113

Enhancing Remineralization of Subsurface Enamel Lesions with Functionalized β-TCP

Robert L. Karlinsey*[*], *Allen C. Mackey, Emily R. Walker and Katherine E. Frederick
Indiana Nanotech, Indianapolis, IN, US

Research Summary

In an attempt to further the anticaries benefits afforded by fluoride, we have developed a unique functionalized tricalcium phosphate (fTCP) and explored its remineralization potential in several dental models. fTCP was found to be stable in NaF aqueous solutions of 500, 950, and 5000 ppm F, and is the first form of bioavailable calcium that is stable with fluoride in an aqueous and single-compartment environment. Using an *in vitro* pH cycling regimen emulating actual events transpiring in the oral cavity, we demonstrated that the baseline remineralization of softened enamel with topical solutions of 500, 950, and 5000 ppm F can be substantially improved when combined with fTCP. Using both Vickers and Knoops microhardness measurements, the reversal of white-spot lesions was attributed to both surface and bulk remineralization effects with fTCP providing statistical improvement over the base fluoride solution. However, in a single-step dental model, fTCP was found to limit fluoride uptake. This result was inconsistent with multiple-step regimens and previously reported work, but furthers our understanding of the influence fTCP has on enamel. Because fTCP does not compromise fluoride bioavailability and is not limited to boosting remineralization through elevating fluoride uptake alone, we explored the role fTCP has with enamel tissue and fluoride. Based on our results, we conclude fTCP intertwines synergistically with both fluoride and the enamel tissue in a sophisticated manner that produces superior remineralization of the subsurface lesion relative to fluoride alone.

[*] Corresponding author: E-mail: rkarlins@gmail.com; Ph: 317-278-7892.

In: Biochemistry Researcher Biographical Sketches ...
Editors: J. T. Walker and C. M. Jones
ISBN: 978-1-62081-308-9

Chapter 114

Translation of Emerging Hydrogel Therapies: The Role of Metrology

***Melissa L. Mather*[1,2] *and Paul E. Tomlins*[1]**
[1]Materials Division, National Physical Laboratory, Teddington, Middlesex, UK
[2]Electrical Systems and Optics Research Division, University of Nottingham, Nottingham, UK

Research Summary

Traditionally hydrogels have been widely used in medicine for sight correction, as coupling agents, barrier coatings and vehicles for drug delivery. Such materials are relatively simple in terms of their design and performance. This situation is rapidly changing as hydrogels begin to find a use in emerging healthcare solutions particularly in regenerative medicine and advanced drug delivery. Hydrogels capable of stimuli responsiveness, self-assembly, and molecular selectivity can now be produced for therapeutic applications owing to advances in hydrogel synthesis.

Despite the growth of potential hydrogel based therapies, the route to market for new products is constrained by the need to develop an appropriate regulatory and metrological framework under which they will be assessed. This chapter will outline the role that metrology has to play in the regulation of emerging hydrogel-based therapies and detail possible strategies for assessing their safety and effectiveness. It will be concluded that there is a need to improve current metrology in relation to hydrogels in order to expedite their translation into clinical practice.

In: Biochemistry Researcher Biographical Sketches ... ISBN: 978-1-62081-308-9
Editors: J. T. Walker and C. M. Jones

Chapter 115

Metal Ion Release from the Base Co▪Cr▪Mo, Ni▪Cr, and Noble Au▪Pt Dental Alloy into the Buffered Solutions of Different Composition and pH Value

A. Celebic*[1], B. Momcilovic*[2] *and S. Miko*[3]**
[1]School of Dental Medicine, University of Zagreb
[2]Institute for Research and Development of Sustainable Eco Systems.
[3]Institute of Geology, University of Zagreb, Zagreb, Croatia

Research Summary

Objectives: The prevailing biological conditions in the human oral cavity induce the release of trace element (TE) metal cations from the dental casting alloys. We studied the release of metal ions (Al, Ag, Au, Ca, Cd, Co, Cr, Cu, Mg, Mo, Ni, Pd, Pt, Ti, and Zn) from three most common dental casting alloys, two base and one noble: Co▪Cr▪Mo, Ni▪Cr, and Au▪Pt alloy into three buffered biological solutions simulating the composition and pH of respective saliva, acidity, and dental plaque.

Methods: The commercially available Co▪Cr▪Mo, Ni▪Cr and Au▪Pt alloy specimens were soaked in pH 6.0 phosphate buffer (*Saliva*), 3.5 pH phosphate buffer (*Acid*) and pH 3.5 mixture of lactic, formic and acetic acid (dentobacterial *Plaque*), and incubated at 37 ^{0}C for 1, 2, 3, 4, 5, 6, 7, 14, 21, and 30 days. Six samples (n = 6) of every solution were prepared for every time period. Inductively coupled plasma atomic emission spectroscopy was used for analysis of the released elements.

Results: The concentration of the trace elements (TE) released from soaked dental metal alloys in various solutions of different pH are expressed in $\mu g \cdot L^{-1}$ as Mean values (SD). Average detectable amounts of TE from the Co▪Cr▪Mo dental alloy were released into: *Plaque* Co 502 (412), *Acid* Cr 589 (84), *Plaque* Fe 180 (43), *Acid* Ni 80 (96), and *Acid* Zn 87 (56). The manufacturers did not indicate the presence of Fe, Ni, and Zn. Average detectable amounts of TE from the Ni▪Cr dental alloy were released into: *Acid* Co 347 (394), *Saliva* Cr

397 (491), *Saliva* Fe 248 (257), *Saliva* Ni 542 (669), and Plaque Zn 96 (51). The manufacturers did not indicate the presence of Co and Zn. Average detectable amounts of TE from the Au▪Pt dental alloy were released into: *Acid* Cr 895 (14), *Plaque* Cu 113 (79), *Plaque* Zn 207.4 (25.3), and *Acid* Fe 150 (43). The manufacturers did not indicate the presence of Cr and Fe. ANOVA revealed the significant effect of alloy composition, the respective buffer solution *(Saliva, Acid, Plaque)*, time interval *(1 to 30 days)*, and interaction on ion release ($P<0.001$ for every compared parameter).

Conclusion: There was a great within the buffer solition variability in metal ion release from the same dental alloy indicating their inhomogeneity. The base metal alloys virtually released more ions than the Au▪Pt alloy, however their effective surface area was larger than that of noble alloy. The ion release from an alloy doesn't necessarily correlate with the abundance of the element in the alloy. Rather, there is a selective dissolution so that the elements that are present in alloys only in traces can be released from them in larger amounts. The concentration of the major essential TE of Cu, Fe and Zn released from base dental alloys was below the Reccomended Dietary Allowances (RDA's), whereas the leaching concentrations of the allergogenic Co, Cr, and Ni were substantial. Considerable amounts of Ni was released from Ni▪Co dental alloy, whereas undeclared Co from the Ni▪Cr alloy and undeclared Ni from the Co▪Cr▪Mo alloy may have both local and systemic, especially allergognic, adverse effects. The undeclared chromium from noble Au▪Pt dental alloy appears to be responsible for the contact allergy thus far attributed to the gold.

In: Biochemistry Researcher Biographical Sketches … ISBN: 978-1-62081-308-9
Editors: J. T. Walker and C. M. Jones

Chapter 116

Co-Cr-Mo Alloy Surface Features and Composition Prior and after Mechanical Polishing and Corrosion in Fluids Simulating Oral Conditions

Asja Celebic, Vesna Svetlicic[1], Jadranka Malina[2] and Boris Klaic[3]
Department of Prosthodontics, School of Dental Medicine, University of Zagreb, Croatia
[1]R.Boskovic Institute, Zagreb, Croatia
[2]Faculty of Metallurgy, Zagreb, Croatia
[3]Polyclinic of Dentistry Perkovceva, Zagreb, Croatia

Research Summary

Introduction: It has been well-documented that metal ions are released from all dental alloys, which may be of a clinical concern and may be a potential health problem. Corrosion behaviour in oral fluids is strongly influenced by microstructural characteristics and surface properties of dental alloy.

Aim of the Study: To examine the surface of CoCrMo alloy prior and after exposure into different solutions simulating oral conditions, as well as prior and after different mechanical polishing procedures.

Materials and Methods: The cast and electropolished specimens of CoCrMo alloy (EP) (Wironit®, extra hard) were examined by SEM with EDS and by AFM prior and after immersion into two solutions simulating saliva and dentobacterial plaque for 30 days at 37 °C. The effect of polishing at the surface of CoCrMo alloys has also been studied. After electropolishing procedures (EP, 6 samples) another samples were further mechanically polished using green rubber discs and afterwards a high shine polishing paste and a rotating black brush (BB, 6 samples), or using green rubber discs, a high shine polishing paste and a rotating deer leather brush (DL, 6 samples).

Results: The AFM analysis provided usefull information about the surface morphology. AFM imaging revealed that the surface of EP specimens prior the contact with corroding solutions was characterized by clusters of nanocrystallites forming islands emerging from the smooth undulating surface. The islands' height was in the range of 90-150 nm. Besides undulating surface and islands of nanocrystallites, some high conical crystallites and small grains can also be observed. The undulating surface was twice richer with Co than Cr, while in the clusters of nanocrystallites (islands) Co was almost equal to Cr, which is different from the declared alloy's composition. However, some cristallytes were dissolved after corrosion, but some new phases (dense sprinkles of very small crystallites) appeared. The respective alloy peaks in the EDS spectrum revealed that these phases were a result of a corrosion - primarily of metal dissolution and salt formation. The presence of some elements from the corroding solution was registered at the alloy surface, as well. The effect of mechanical polishing was clearly evident at CoCrMo surface. The clusters of nanocrystallites forming islands (emerging from the undulated surface) were smoothened and/or thorn away by further mechanical polishing, while the islands' perimetar was unaltered. Mechanical polishing also produced scratches in both, BB and DL samples. The scratches were observed not only at the undulating surface, but also at the smoothed islands. The Z range (distance from the highest to the lowest peak) of EP samples revealed significantly higher values than that of BB and DL ($p<0.01$). The highest RMS and the Ra values were also recorded in the EP samples, but not significantly different ($p>0.05$), compared to the mechanically polished BB and DL samples.

Conclusions: AFM imaging revealed at least four different phases at the surface of CoCrMo electropolished specimens. The most dominant phases were: undulating surface and clusters of nanocrystalites emerging from the undulating surface forming islands, but some high conical crystallites and some small grain-like crystallites have also been observed. Mechanical polishing of EP samples smoothed islands of nanocrystallites, but also formed scratches at the surface, which were more prominent in DL than BB samples. During the corrosion process of EP samples probably metal salts (anions from corroding solutions) were binded at the surface as a new phase. Corrosion behaviour of CoCrMo alloys is clearly affected by pH and composition of the fluids. It is obvious that some complex mechanisms involving reactions between ionic species are dominating the interfacial processes, which in turn affect the morphology and the composition of the phases formed.

In: Biochemistry Researcher Biographical Sketches ... ISBN: 978-1-62081-308-9
Editors: J. T. Walker and C. M. Jones

Chapter 117

Intracellular Delivery of Gold Nanoparticles: Applications in Nanomedicine

Deendayal Mandal*
Department of Biomedical and Pharmaceutical Sciences,
University of Rhode Island, Kingston, US

Research Summary

Nanotechnology has enormous potential in biomedical research including disease diagnosis and therapy. Inorganic nanoparticles have been adopted for biomedical research due to their unique optical and physical properties. Compared to conventional materials, inorganic nanomaterials have several advantages such as simple preparative process and precise control over their shape, composition and size. A major aim of medicine is the early and precise diagnosis of clinical conditions, providing an efficient treatment without secondary effects. Nanoparticle-based technologies have demonstrated especially high potential for medical purposes, ranging from diagnosing diseases to providing novel therapies. Among the metal nanoparticles, gold nanoparticles have proven to be powerful tools in various nanomedicinal applications because of their photo-optical properties and biocompatibility. However, the existing nanoparticle-based technologies must overcome several challenges, including selective nanoparticle delivery, potential toxicity, imaging of nanoparticles and therapeutic efficacy. This chapter describes a brief overview of the recent research on delivery of gold nanoparticles inside the cells and their potential applications in nanomedicine.

* Corresponding author: Department of Biomedical and Pharmaceutical Sciences, 41 Lower College Road, University of Rhode Island, Kingston, RI 02881 (USA), Email: ddmandal@yahoo.com, Ph: +1 401 874 4221, Fax: +1 401 874 5787.

In: Biochemistry Researcher Biographical Sketches ... ISBN: 978-1-62081-308-9
Editors: J. T. Walker and C. M. Jones

Chapter 118

A Commentary on Neural Tissue Engineering in the Central Nervous System – Interfacing a Lesion

David R. Nisbet, James A. Bourne and John S. Forsythe
Department of Materials Engineering, Monash University, Melbourne, Australia

Research Summary

Damage to the Central Nervous System (CNS) from disease or injury remains a serious problem that affects millions of individuals worldwide. Regeneration across a lesion in the adult mammalian CNS is autonomically inhibited, thereby presenting substantial challenges for clinical treatment strategies. Current methods employed such as autografts, xenografts and allografts have limited success due to the complex nature of CNS injury pathophysiology. As a result, there is no effective treatment available to establish repair of damaged neurones within the CNS, with patients typically being administered a combination of drugs and rehabilitation in attempts to maintain and restore some functionality. Recent research has explored the potential of synthetic scaffolds to fabricate biomimetic regenerative microenviroments for nerves as an alternative treatment option. This commentary addresses the importance of scaffold design in enhancing the regeneration capacity of the adult CNS; focusing on past innovative strategies and current problems. Future trends towards fabricating scaffolds to achieve superior regenerative outcomes will also be highlighted.

In: Biochemistry Researcher Biographical Sketches … ISBN: 978-1-62081-308-9
Editors: J. T. Walker and C. M. Jones

Chapter 119

Biosurfactants and Their Uses in the Petroleum Industry and Hydrocarbon Pollution Remediation

N. L. Olivera[*] and M. L. Nievas
Centro Nacional Patagónico (CONICET), Chubut, Argentina

Research Summary

Biosurfactants are amphipathic molecules which have the capability to reduce the medium surface tension. At low concentrations, surfactants in aqueous solution exist as monomers. As the surfactant concentration increases, the medium surface tension decreases. Above a threshold concentration known as the critical micelle concentration (CMC), biosurfactant molecules form micelles with a hydrophobic core and a hydrophilic surface. CMC values are characteristic of each biosurfactant, depending on the chemical structure and composition of the biosurfactant as well as the solution conditions (e.g. temperature, ionic strength, additives). There is an important diversity in chemical structures and functions among natural surfactants. A distinction between low and high-molecular-weight biosurfactants is accepted, the former are generally glycolipids or lipopeptides with surface activity, while the latter are exocellular polymeric substances with important emulsifying properties. This chapter focuses on microbial biosurfactants that on the context of degradation of lipophilic compounds, such as hydrocarbons, may enhance their bioavailability to microorganisms, either by increasing the apparent hydrocarbon solubility in the aqueous phase or by increasing the contact surface area due to emulsification. In this regard, different biosurfactants have been applied to enhance the biodegradation of oily residues and to fight against oil pollution. In addition, biosurfactants possess the advantage of being biodegradable molecules which encourages its use in the development of bioremediation strategies. In the petroleum industry, biosurfactants also find application in MEOR (microbial enhanced oil recovery) improving hydrocarbon mobilization and consequently the recovery of crude oil from reservoirs.

[*] Tel. +54-2965-451024; Fax +54-2965-451543. E-mail addresses: *olivera@cenpat.edu.ar*. nievas@cenpat.edu.ar.

In: Biochemistry Researcher Biographical Sketches ... ISBN: 978-1-62081-308-9
Editors: J. T. Walker and C. M. Jones

Chapter 120

Detergents in Molecular Biology: DNA Extraction and Purification

Christian Alberto García-Sepúlveda*, Sandra Elizabeth Guerra-Palomares and Diana Lorena Alvarado-Hernández
Laboratorio de Genómica Viral y Humana, Facultad de Medicina, Universidad Autónoma de San Luis Potosí. San Luis Potosí, México

Research Summary

Over the last 20 years, epidemiologist have shifted their attention from the analysis of phenotypic features to evaluating the way genetic traits determine disease causation and severity. This trend has clearly benefitted from the coming of age of versatile molecular based techniques such as the Polymerase Chain Reaction (PCR) and the now widespread availability of high-throughput molecular screening assays. In this context, deoxyribonucleic acid (DNA) has rapidly become the biospecimen of choice as it is readily obtained, transported, processed and stored. Both molecular epidemiology and genomic characterization studies require the typing of large numbers of specimens in order to achieve statistical significance and enormous advances, such as microarray technology, have resulted as a consequence of this. Nevertheless a usually overlooked yet crucial aspect of nucleic acid based typing is the DNA extraction process. DNA extraction is the process by which nucleic acids are retrieved from their biological compartments and deprived of proteins, salts and other substances that might hinder their analysis. In their natural state nucleic acids are compartmentalized inside the cell nucleus or mitochondrion and protected by proteins and a lipid envelope. In order to extract these nucleic acids cells must be burst open and the contaminating proteins and lipids eliminated. Detergents of essentially every type have been

* Corresponding author: Christian A. García Sepúlveda, MD PhD, Laboratorio de Genómica Viral y Humana, Facultad de Medicina Universidad Autónoma de San Luis Potosí, Avenida Venustiano Carranza #2405, Colonia Filtros las Lomas, CP 78210, San Luis Potosí, México, +52(444) 826-2344 extension 568, +52 (444) 817-6976, christian.garcia@uaslp.mx

used to disaggregate the lipid components of cell membranes and are common components of essentially all DNA extraction methods. In this chapter a brief historical review of the different DNA extraction techniques that have been developed is made followed by a detailed presentation of the most relevant applications of detergent-based DNA extraction techniques. Finally, we describe our experience at implementing and optimizing a simple and low-cost laundry-detergent-based DNA extraction method for the extraction of high-quality human genomic DNA from peripheral blood samples and the many uses that such material has been put to use for.

In: Biochemistry Researcher Biographical Sketches ... ISBN: 978-1-62081-308-9
Editors: J. T. Walker and C. M. Jones

Chapter 121

Carbohydrases in Detergents

Piamsook Pongsawasdi[1] and Shuichiro Murakami[2]
[1]Department of Biochemistry, Faculty of Science, Chulalongkorn University, Thailand
[2]Department of Agricultural Chemistry, Faculty of Agriculture, Meiji University, Japan

Research Summary

The chapter starts with an introduction on detergents and the detergent industry. The history and development of enzyme usage as an active ingredient in detergents for laundry, dishes and other household products are listed. The importance and the main characteristics of enzymes that are required so as to be suitable for use in detergents are described. The properties and functions of hydrolases (proteases, carbohydrases and lipases) that are beneficial for the detergent industry are overviewed. Then the carbohydrases that have specific functions in detergents are focused upon, especially α-amylase, cellulase and mannanase. For each enzyme type, the catalytic reaction, sources and production of the enzyme, enzyme specificity, biochemical characteristics, function and development of the enzyme for desired properties through protein engineering, are discussed. The work of our group on a new highly alkaline, thermostable α-amylase that is resistant to certain surfactant and bleaching reagents commomly used as additives to detergents, and so has a good potential in the detergent industry, is presented. The commercial carbohydrases that have been used in detergent industry are reviewed. Finally, this chapter ends with reviewing some of the future prospects on the use of enzymes in detergents.

In: Biochemistry Researcher Biographical Sketches ... ISBN: 978-1-62081-308-9
Editors: J. T. Walker and C. M. Jones

Chapter 122

The Influence of Different Detergents on the Preservation of Discrete Membrane Microdomains: A Comparative Ultrastructural Study on Whole-Mounted Colorectal Cancer Cells

Kristina A. Jahn**[*]*, *Yingying Su*, *Tessa Smissaert van de Haere*, *Iris Benjamins and Filip Braet***
Australian Centre for Microscopy & Microanalysis,
The University of Sydney, Australia

Research Summary

The cellular plasma membrane is partially resistant to solubilisation with mild and non-denaturing detergents allowing detergent-resistant membranes (DRMs) to be isolated and biochemically analysed. The composition of DRMs however varies depending on the detergent applied. We previously developed a novel method to visualise DRMs on whole-mount cancer cells in which the soluble fraction of the plasma membrane is removed and the cytoskeleton is fixed and prepared for scanning electron microscopy. This method is less destructive than many others commonly used to study DRMs and it enables the direct visualisation of intact, cytoskeleton-associated DRMs. In this chapter we compare five commonly used detergents (TX-100, Brij96, CHAPS, Lubrol WX and OG) with respect to the achieved membrane solubilisation at different concentrations. Our results indicate that TX-100, Brij96 and CHAPS isolate similar DRMs whereas OG fully solubilised the plasma membrane. Interestingly, Lubrol WX poorly extracted cells even at high concentrations. This

[*] Corresponding author: Australian Centre for Microscopy & Microanalysis (ACMM), Madsen Building F09, The University of Sydney, NSW, 2006, Australia. Tel: + 61 2 9351 7551; Fax: + 61 2 9351 7682; E-mail: kristina.jahn@sydney.edu.au.

is the first ultrastructural study that systematically describes and visualises the effects of different detergents on the preservation of the cytoskeleton and DRMs on whole-mounted cells. This approach delivers new and valuable ultrastructural information at the nanometer scale that did not exist to date.

In: Biochemistry Researcher Biographical Sketches ... ISBN: 978-1-62081-308-9
Editors: J. T. Walker and C. M. Jones

Chapter 123

Φ-Sulfo Fatty Methyl Ester Sulfonates (Φ-MES): A New Anionic Surfactant

León Cohen[1*], Fernando Soto[1], Francisco Trujillo[1], David W. Roberts[2] and Claudio Pratesi[3]

[1] Escuela Politécnica Superior de Algeciras.
Universidad de Cádiz, Spain
[2] Liverpool John Moores University
[3] Sasol Italy S.p.A

Research Summary

Φ-Sulfo Fatty Methyl Esters sulfonates (Φ- MES) are new anionic surfactants obtained via sulfoxidation of fatty acid methyl esters (FAME) with SO2, O2, and ultraviolet light of appropriate wavelength. The designation of Φ refers to the random positioning of SO3 in the alkyl chain.

In this work we summarize the most relevant results of our research started fifteen years ago, on the synthesis, separation, analysis and performance of Φ-sulfo fatty methyl ester sulfonates known as Φ-MES, and we update our last findings . We have to point out that we are for the time being, the only research group in the world to have published our research on Φ-MES.

This work describes the optimum batch conditions for the sulfoxidation of FAME and a reaction mechanism is proposed. The paper depicts an improved workup for the separation of reaction products from non reacted methyl ester and the GC-MS analysis of Φ- sulfo fatty methyl ester sulfonate is shown. Besides, an interpretation of conversion and selectivity of sulfoxydation reaction is given.

Finally, performance of water solutions based on sulfoxylated methyl ester of palmitic acid (Φ-MES C16) have been studied and compared to two leading types of surfactants used today: linear alkylbezene sulfonate (LAS) secondary alkane sulfonate (SAS) and to α-sulfo

* Correspondance author: E-mail: *leon.cohen@uca.es* Ramón Puyol s/n 11202 - Algeciras (Spain).

fatty methyl ester sulfonate (α-MES) with regard to solubility, performance and skin compatibilty. The experimental results obtained indicate that Φ-MES can be regarded as a potential component of detergent formulations and most likely of body care products.

In: Biochemistry Researcher Biographical Sketches ... ISBN: 978-1-62081-308-9
Editors: J. T. Walker and C. M. Jones

Chapter 124

Acacia Caven (Mol.) Molina Pollen Proteases: Application to the Peptide Synthesis and to Laundry Detergents

***Cristina Barcia*[1], *Evelina Quiroga*[1,3,*], *Carlos Ardanaz*[2], *Gustavo Quiroga*[1] and *Sonia Barberis*[1,3,]**
[1]Laboratorio de Bromatología
[2]Área de Química Orgánica. Facultad de Química, Bioquímica y Farmacia. Universidad Nacional de San Luis. Chacabuco y Pedernera (5700) San Luis, Argentina
[3]Instituto de Física Aplicada (INFAP).
Universidad Nacional de San Luis, Argentina

Research Summary

It is known that the proteases have applications in several industrial processes such us leather processing, laundry detergents, producing of protein hydrolysates and food processing, as well as in the peptide synthesis in non conventional media. The application of proteases as catalyst of short oligopeptides in aqueous-organic media, have received a great deal attention as a viable alternative to chemical approach because of their remarkable characteristics. On the other hand, alkaline proteases have also been used to improve the cleaning efficiency of detergents. Detergent enzymes account for about 30% of the total worldwide enzyme production and represent one of the largest and most successful applications of modern industrial biotechnology. The aim of this work was to study the performance of proteolytic enzymes of Acacia caven (Mol.) Molina pollen for its potential application as an additive in various laundry detergents formulations and as catalyst of the peptide synthesis in aqueous-organic media. Pollen grains (35 mg/ml) were suspended in 0.1M Tris-HCl buffer pH 7.4 and slowly shaken for 2 h at 25° C. Then, the slurry was centrifugated for 30 min at 8000 rpm and the supernatant (crude enzyme extract, CE) was tested in protein content (Bradford's method) and proteolytic activity (using BAPNA and Z-

[*] E-mail address: equiroga@unsl.edu.ar. sbarberi@unsl.edu.ar.

Ala-pNO as substrates). A partial characterization of Acacia caven CE was carried out: enzyme extract displayed maximum proteolytic activity at pH 8 and 35-40º C; it showed remarkable thermal stability after 1.5 h at 25-40º C but it decreased as long as temperature increased to 60º C. On the other hand, the enzyme extract was incubated with different surfactants and commercial laundry detergents at 25-60° C during 30 min and 1h; and it showed high stability and compatibility with them. The peptide synthesis catalyzed by Acacia caven CE was carried out in a mixture of 0.1M Tris-HCl buffer pH 8.5 and ethyl acetate (50:50 ratio) at 37° C using 2-mercaptoethanol as activator and TEA as neutralizing agent of the amino component (Phe-OMe.HCl). Carboxylic components were selected in base of the highest preference of CE. The identification of synthesized peptide products was carried out by HPLC-MS. According to the obtained results, this work contributes with a new variety of phytoprotease useful as catalyst of the peptide synthesis and as additive of laundry detergents.

In: Biochemistry Researcher Biographical Sketches ... ISBN: 978-1-62081-308-9
Editors: J. T. Walker and C. M. Jones

Chapter 125

Pharmacogenetics of Glucocorticoid Response

***Raffaella Franca*[1,3], *Sara De Iudicibus*[2,3] *and Giuliana Decorti*[3*]**
[1]Institute for Maternal and Child Health IRCCS Burlo Garofolo, Trieste,
[2]Department of Reproduction and Development, University of Trieste, Trieste, Italy
[3]Department of Life Sciences, University of Trieste, Trieste, Italy

Research Summary

Glucocorticoids (GC) are first-line treatment in a number of inflammatory, autoimmune and neoplastic diseases. Despite their extensive therapeutic use and the proven effectiveness, considerable clinical evidence of wide inter-individual differences in GC efficacy and side effects among patients has been reported. In recent years, a detailed knowledge of the GC mechanism of action and of the genetic variants affecting GC activity at the molecular level has arisen from several studies. The glucocorticoid receptor (GR) is crucial for the effects of GCs: mutations in the GR gene (NR3C1) are the primary cause of a rare, inherited form of GC resistance; in addition, several polymorphisms of this gene have been described and associated with GC response and toxicity. However, the GR is not self-standing in the cell and the receptor-mediated functions are the result of a complex interplay of GR and many other cellular partners. The latter comprise several chaperonins of the large cooperative hetero-oligomeric complex that binds the hormone-free GR in the cytosol, and several factors involved in the transcriptional machinery and chromatin remodelling, that are critical for the hormonal control of target genes transcription in the nucleus. Polymorphisms in genes involved in the pharmacokinetics and/or pharmacodynamics of these hormones have also been suggested as other possible candidates of interest that could play a role in the observed inter-individual differences, because of their influence on drug disposition. The best characterized example is the drug efflux pump P-glycoprotein (P-gp), a membrane transporter that acts lowering GC intracellular concentration. This protein is encoded by the

[*] Corresponding author: Department of Life Sciences, University of Trieste, Via L. Giorgieri n° 7, I-34127 Trieste, Italy, Tel. 39 040 5587949, Fax. 39 040 5587838, E-mail. decorti@units.it.

ABCB1/MDR1 gene that presents different known polymorphic sites. Furthermore, variants in the principal effectors of GCs (e.g: cytokines and their regulators) have also to be taken into account for a comprehensive evaluation of the great variability in GC response. The goal of this review is to emphasize the current knowledge on this topic and to underlie the role of genetics in predicting GC clinical response. The ambitious prospective of pharmacogenomic studies is to achieve an individualized therapy, giving to physicians a reliable tool based on the patient genetic profile.

In: Biochemistry Researcher Biographical Sketches ... ISBN: 978-1-62081-308-9
Editors: J. T. Walker and C. M. Jones

Chapter 126

The Glucocorticoid Receptor Signaling in Cancer and Inflammation

E. Kassi and P. Moutsatsou*

[1]Laboratory of Clinical Biochemistry, Attikon University Hospital, Athens
[2]Laboratory of Biological Chemistry, Medical School, University of Athens, Goudi, Athens

Research Summary

Glucocorticoids exert pleiotropic effects in many systems of human body, including the immune system, bone, liver, brain, cardiovascular system, as well as, breast and prostate. Glucocorticoids are provided as anticancer therapy in haematological tumors and are used as co-medication with chemotherapy in prostate cancer and breast cancer. Glucocorticoids are among the most important anti-inflammatory agents, however continued exposure of bone to these steroids results in osteoporosis. In this review we provide a detailed description of the structure and function of glucocorticoid receptor, the cornerstone molecule mediating the glucocorticoids' cellular effects, the interaction of glucocorticoid receptor with other signaling molecules such as mitogen activated protein kinases, the transcription factors AP-1 and NF-kB and the mitochondrial genome. We summarize recent data on how glucocorticoid receptor-mediating signaling affects inflammatory processes as well as breast and prostate cancer-related signal transduction pathways via regulation of specific genes including growth factors, cytokines, transcription factors, mitogen activated protein kinases, oncogenes as well as apoptosis and proliferation-related processes known to play key roles in cellular/tissue functions.

* Corresponding author: Laboratory of Biological Chemistry, Medical School, University of Athens, 75 Mikras Asias Street, GR 11527, Goudi, Athens, Greece Tel/Fax: 00302107462682; E-mail: pmoutsatsou@med.uoa.gr.

In: Biochemistry Researcher Biographical Sketches ...
Editors: J. T. Walker and C. M. Jones

ISBN: 978-1-62081-308-9

Chapter 127

Glucocorticoids – Age Specific Impact on Mitochondrial Energy Homeostasis

Jignesh D. Pandya[1*] and Surendra S. Katyare[2]
[1]Spinal Cord & Brain Injury Research Center (SCoBIRC), University of Kentucky, Kentucky, US
[2]Department of Biochemistry, Faculty of Science, The Maharaja Sayajirao University of Baroda, Vadodara, India

Research Summary

The *Zona Fasciculata* region of adrenal cortex synthesizes cortisol- a glucocorticoid group of steroid hormones that stimulates "fight and flight responses" to battle against stress condition. The primary function of glucocorticoids is to supply high energy level to body during stress condition. The endocrine secretions of glucocorticoids are highly regulated by hypothalamus-pituitary-adrenal (HPA) axis. Factors like diurnal variation, age, sex and season have impact on glucocorticoids secretion. Wide application of glucocorticoids as therapeutic agents has been appreciated due to their anti-inflammatory and immuno-suppressive properties. The diagnostic and therapeutic application as single or multiple dosages of glucocorticoids has been given for various conditions like chronic lung diseases (CLD) in extremely low birth weight infants, bronchopulmonary dysplasia (BPD), respiratory distress syndrome (RDS) in fetuses and metastasis of brain, spine and meningus. Although, a synthetic glucocorticoid is believed to be safer and to have no side-effects on above conditions; when given at higher pharmacological doses, they may pose a risk for neuro-developmental processes. Our research studies on developing animals (2 week to 5 week group) and adult (8 week) have specifically shown age and brain specific alterations in mitochondrial structure and function that directly affects the energy homeostasis during these developmental periods after chronic treatment of corticosterone (major glucocorticoid of rat)

* Corresponding author: Research Scientist II, Spinal Cord & Brain Injury Research Center (SCoBIRC), University of Kentucky, B436, BBSRB, 741 South Limestone Street, Lexington, KY 40536-0509, Phone: 859-323-4682, Fax: 859-257-5737, E-mail: Jignesh.Pandya@uky.edu.

and dexamethasone (synthetic glucocorticoid). Our research work also cautions against extensive and repeated use of glucocorticoids due to their high impact on mitochondrial energy metabolism in the brain and the liver during developmental and adult age. This comprehensive review article will cover the history, pharmacology, specific effects of glucocorticoids during development and conclusive remark observed by us and other research laboratories.

In: Biochemistry Researcher Biographical Sketches ... ISBN: 978-1-62081-308-9
Editors: J. T. Walker and C. M. Jones

Chapter 128

Glucocorticoids, Stress and Cancer: New Insights in Immune System and Tumour Cell Regulations

Luciana Romina Frick[1,*], Maximilano Rapanelli[2], Graciela Alicia Cremaschi[1] and Ana María Genaro[1]

[1]Centro de Estudios Farmacológicos y Botánicos, Consejo Nacional de Investigaciones Científicas y Técnicas, Primera Cátedra de Farmacología, Universidad de Buenos Aires, Argentina

[2]Instituto de Biología y Medicina Experimental, Consejo Nacional de Investigaciones Científicas y Técnicas, Buenos Aires, Argentina

Research Summary

Glucocorticoids and catecholamines are considered the main stress hormones, which are produced after the activation of the hypothalamic-pituitary adrenal axis (HPA) and the autonomic nervous system (ANS) induced by an aversive stimulus. Interestingly, exposure to stressful situations has been associated with alterations in the incidence, progression and outcome of neoplastic processes. In addition, corticoids are often co-administered in cancer patients as adjuvant treatment to chemo and radiotherapy. Cortisol has been proposed as one of the hormones responsible of these alterations, in part given that stress and glucocorticoids alter different components of the immune system such as lymphocyte proliferation, cytotoxic and fagocytic activity, antibody and cytokine production, and redistribution of immune cells among the different lymphoid organs. In addition, glucocorticoids have direct effects on cancer cells, for example, they modulate the expression of genes related to chemotherapy resistance, survival and apoptosis evasion in certain tumours. Moreover, some oncogenic virus such as human papilloma, hepatitis and Epstein-Barr viruses are susceptible to be regulated by glucocorticoids. Here we review all these items and analyse the recent advances

* Corresponding autor: lfrick@fmed.uba.ar.

in this field obtained using animal models and clinical studies. Also, we discuss the novel discovery of the involvement of hypothalamic-pituitary thyroid (HPT) axis in these processes, and the regulation exerted by the HPA on thyroid hormones.

In: Biochemistry Researcher Biographical Sketches … ISBN: 978-1-62081-308-9
Editors: J. T. Walker and C. M. Jones

Chapter 129

Hypothalamic Obesity: A Pivotal Role of Endogenous Glucocorticoid

***Eduardo Spinedi*[1,*]*, Andres Giovambattista*[1] *and Rolf C. Gaillard*[2]**
[1]Neuroendocrine Unit, Multidisciplinary Institute on Cell Biology (IMBICE, CONICET-CICPBA), Argentina
[2]Division of Endocrinology, Diabetology & Metabolism, University Hospital (CHUV), Lausanne, Switzerland

Research Summary

It is known that the neonatal treatment of rats with monosodium L-glutamate (MSG) induces several metabolic and hormonal abnormalities, resulting in enhanced adiposity and hyperleptinemia, among others. Our studies explored the consequences of MSG-induced enhanced hypothalamo-pituitary-adrenal (HPA) axis function and chronic hyperleptinemia on the adrenal sensitivity to the inhibitory effect of exogenous leptin. In addition, in this chapter includes the consequences of adrenal gland manipulations on several characteristics of the hypothalamic obese phenotype. Neonatal male rats treated with MSG or vehicle (controls, CTR) were followed during 150 days in order to study changes observed over development in body weight, food consumption as well as *in vivo* HPA axis and adipocyte functions. During adulthood, adrenal response to adrenocorticotropin (ACTH) was evaluated both *in vitro* and *in vivo* in order to determine the adrenal sensitivity to the inhibitory effect of leptin. For this purpose, sham operated as well as CTR and MSG rats with bilateral adrenal enucleation (AE) were used. Finally, by studying AE and bilaterally adrenalectomized (ADX) adult rats, either receiving corticosterone replacement therapy or not, we addressed modifications on both neuroendocrine-metabolic functions and adiposity depots. Our results indicate that: 1) between 30 and 150 days of age, MSG animals developed hypophagia, accompanied by arrest in body weight gain, and concomitant enhanced basal levels of all HPA axis components and of leptin; 2) adrenals from of 150 day-old MSG rats displayed an *in vitro* adrenocortical

* Corresponding author: Neuroendocrine Unit, IMBICE, PO Box 403, 1900 La Plata, ARGENTINA, FAX: +54.221.4253320, E-mail: spinedi@imbice.org.ar.

hyperresponse to ACTH stimulation as well as an adrenal refractoriness to the physiological inhibitory effect of leptin on ACTH-stimulated glucocorticoid output; and 3) bilateral adrenal enucleation (AE) in adult MSG-treated rats transiently reversed the MSG-induced hyperleptinemia, restoring normal leptin levels as well as a normal adrenal sensitivity to the inhibitory effect of leptin. While 21 days of either ADX or AE blunted enlarged adiposity mass in MSG rats, corticosterone replacement therapy in ADX MSG rats normalized adiposity. Our data indicate that adrenal exposure to the chronically high plasma leptin levels observed in MSG rats is involved in the loss of the inhibitory regulatory effect of leptin at the adrenal level, being therefore, at least in part, responsible for the increased total and free glucocorticoid peripheral levels found in adult MSG rats. Furthermore, this study strongly suggests that the adrenal over-function, frequently associated with different phenotypes of obesity, could be due to an adrenal resistance to the leptin negative regulation.

In: Biochemistry Researcher Biographical Sketches …
Editors: J. T. Walker and C. M. Jones
ISBN: 978-1-62081-308-9

Chapter 130

The Role of Nitric Oxide – Redox Imbalance in Glucocorticoid-Induced Hypertension

***Yi Zhang*[1]* *and Judith A. Whitworth*[2]**
[1]Cardiovascular Research, Canberra Hospital, Canberra, Australia
[2]John Curtin School of Medical Research, Australian National University, Canberra, Australia

Research Summary

Excess glucocorticoid increases blood pressure in humans and animals. It has been assumed that glucocorticoid-induced hypertension is due to glucocorticoid activation of mineralocorticoid receptors with consequent salt and water retention but evidence for this view is lacking. Experimental cortisol-induced hypertension in humans and adrenocorticotrophic hormone (ACTH)-induced hypertension in the rat are not prevented by the mineralocorticoid receptor blocker spironolactone, and synthetic glucocorticoids raise systolic blood pressure without any sodium retention or expansion of plasma volume.

Glucocorticoid-induced hypertension is characterized by nitric oxide (NO) deficiency. Renal nitric oxide synthase (iNOS and eNOS) expression is down-regulated by adrenocorticotrophic hormone (ACTH), corticosterone and the synthetic glucocorticoid dexamethasone. Glucocorticoid-induced hypertension is associated with decreased plasma nitrate/nitrite concentrations in humans and rodents. Direct NO donors (isosorbide dinitrate and DETA/NO) and lipopolysaccharide, a stimulus for iNOS, reverse ACTH-induced hypertension in rodents. Furthermore, the NOS precursor *L*-arginine prevents and/or reverses ACTH- and corticosterone-induced hypertension in rats suggesting a role for the NO system in glucocorticoid-induced hypertension. The pressure lowering effect of *L*-arginine in ACTH

* Corresponding author: Cardiovascular Research, Level 5, Building 10, The Canberra Hospital, Yamba Drive, Garran, ACT 2605, Ph: +61 2 6244 4057, Fax: +61 2 6244 3092 , E-mail: yzhang1962@yahoo.com.

treated rats is reversed by the NOS inhibitor *N*-nitro-*L*-arginine indicating that *L*-arginine is likely to be working through NO generation.

Superoxide scavenges NO and reduces its bioavailability. In rodents, glucocorticoid-induced hypertension is associated with increased oxidative stress. Increased biomarkers of oxidative stress such as plasma F_2-isoprostane concentrations and lucigenin-enhanced chemiluminescence in the aorta were found in ACTH- and dexamethasone-treated hypertensive rats. Further, the superoxide scavenger tempol, and other agents that have antioxidant properties such as *N*-acetylcystein, folic acid, aspirin, losartan, ramipril and alpha-lipoic acid prevent and reverse ACTH- and/or dexamethasone-induced hypertension in the rat. The NADPH oxidase inhibitor apocynin, but not the xanthine oxidase inhibitor allopurinol, prevents and reverses ACTH- and dexamethasone-induced hypertension. Supplementation of sepiapterin or tetrahydrobiopterin has no effect on ACTH- or dexamethasone-induced hypertension in the rat.

These data suggest NO-redox imbalance plays an important role in glucocorticoid-hypertension. This chapter examines the mechanisms of NO-redox imbalance in glucocorticoid-hypertension.

In: Biochemistry Researcher Biographical Sketches ... ISBN: 978-1-62081-308-9
Editors: J. T. Walker and C. M. Jones

Chapter 131

Glucocorticoids Mediate Drug Resistance: Mechanisms and its Clinical Impact

Weimin Fan[1,2] and Meihua Sui[1,3]

[1]Program of Innovative Cancer Therapeutics, First Affiliated Hospital of Zhejiang University College of Medicine, Hangzhou, China;

[2]Department of Pathology and Laboratory Medicine, Medical University of South Carolina, Charleston, US

[3]Center for Bionanoengineering, Department of Chemical and Biological Engineering, Zhejiang University, Hangzhou, China

Research Summary

Drug resistance is one of the major obstacles limiting the success of cancer chemotherapy. Biological mechanisms contributing to drug resistance may be present *de novo* and relate to inherent cellular features or may arise after exposure to anticancer drugs. In this field, the most widely studied phenomenon is multidrug resistance (MDR) that has been linked to overexpression of a membrane associated P-glycoprotein (P-gp). In recent years, both clinical observations and experimental studies suggested that steroid hormones and their receptors might also affect the therapeutic efficacy of antineoplastic drugs. This article will introduce the latest research progresses on the discovery and characterization of the drug resistance mediated by glucocoticoids.

Glucoroticoids are essential steroid hormones required for the maintenance of many physiological and developmental processes. Glucocorticoids such as dexamethasone are widely used as premedication to prevent the side or adverse effects of many chemotherapeutic agents such as paclitaxel and docetaxel. Recent studies in our laboratory and others discovered that glucocorticoids could selectively inhibit paclitaxel-induced apoptosis *in vitro*. Through development of animal models bearing human breast and ovarian xenograft tumors, we demonstrated that pretreatment of dexamethasone significantly attenuated the therapeutic efficacy of paclitaxel *in vivo*. Further investigation indicate that glucocorticoids have little influence on the ability of paclitaxel to induce microtubule bundling and mitotic arrest, but specifically block apoptotic cell death by antagonizing the regulatory effect of paclitaxel on activation of NF-κB/IκB signaling pathway. Since glucocorticoids are routinely used in the clinical application of paclitaxel to prevent hypersensitivity reactions and other adverse effects, the inhibitory action of glucocorticoids on paclitaxel-induced apoptosis also raises a clinically relevant question as to whether the pretreatment with glucocorticoids might interfere with the therapeutic efficacy of paclitaxel and possibly other chemotherapeutic agents.

In: Biochemistry Researcher Biographical Sketches ... ISBN: 978-1-62081-308-9
Editors: J. T. Walker and C. M. Jones

Chapter 132

Glucocorticoids in Perioperative Care

Arman Kahokehr*, Sanket Srinivasa, Tzu Chieh Yu, Tarik Sammour and Andrew G. Hill
Department of Surgery, University of Auckland, Middlemore Hospital, Otahuhu, Auckland, New Zealand

Research Summary

Surgery elicits a stress response of combined endocrine and inflammatory origin, which leads to increased functional demands on various organ systems. This may subsequently contribute to postoperative morbidity. Even after hospitalisation, patients continue to suffer from lack of energy, loss of function and prolonged fatigue. In recent years surgical research has been focused on creating an environment in which surgical stress is minimised and recovery is enhanced. Glucocorticoids are an important class of perioperative drug to achieve this goal.

Glucocorticoids have well-known anti-inflammatory actions and therefore represent a useful peri-operative drug in modern surgery. Their therapeutic benefits go beyond the traditional uses to reducing post-operative nausea and vomiting in adult and paediatric procedures. Recent studies support glucocorticoid administration in a variety of procedures. In controlled and optimised peri-operative conditions, systemic glucocorticoids have been shown to reduce the local and systemic neuro-endocrine response to surgical insult. This translates into post-operative clinical benefits by decreased pain, vomiting, wound related complications and post-operative fatigue (POF). This is an important benefit that goes beyond the half time of the circulating drug to improve long term recovery. Acceleration of recovery with systemic glucocorticoid administration is therefore possible. Surgeons need to be aware of recent developments around glucocorticoid use in surgery.

* Corresponding author: South Auckland Clinical School, PO Box 93311, Auckland, New Zealand, Email: arman.kahokehr@gmail.com, Telephone: +64 9 2760044 extension 4902, Fax: +64 9 2760066.

In: Biochemistry Researcher Biographical Sketches …
Editors: J. T. Walker and C. M. Jones
ISBN: 978-1-62081-308-9

Chapter 133

Glucocorticoid-Mediated Regulation of Tight Junctions in Brain Vascular Endothelium

Malgorzata Burek and Y. Carola Förster*
University Hospital Würzburg, Department of Anaesthesia and Critical Care, Würzburg, Germany

Research Summary

Glucocorticoids (GCs) have anti-inflammatory and immunosuppressive properties. The blood-brain-barrier (BBB) maintains homeostasis of the central nervous system and is disrupted in many pathological disorders. GC treatment improves BBB-properties, and prevents influx of water and harmful substances into the brain. GCs act through genomic and non-genomic mechanisms. The genomic effects are mediated by the glucocorticoid receptor (GR). Ligand bound GR recognizes the glucocorticoid response element (GRE) within the promoter region of genes. It has long been recognized that for modulation of the immune response, GR interferes with other transcription factors such as AP-1 and NFκB. Additionally many co-regulators and co-repressors of GR action have been described. Recent insight shed light on transactivation of certain target genes leading to improved barrier properties in vascular endothelium. In this chapter, we summarize the recent results in the GC-mediated regulation of vascular function with the focus on the BBB. First, we discuss the direct GC targets within junctional proteins, such as occludin, claudins and VE-cadherin. Next, we describe the GC-mediated regulation of vascular function under some pathological conditions, such as multiple sclerosis, inflammation and ischemia. Finally, the issue of GC side effects will be discussed.

* Corresponding author: University of Würzburg, Department of Anaesthesia and Critical Care, Oberduerrbacher Str. 6, 97080 Würzburg, Germany. Tel.: +49 931 20130065; fax: +49 931 20130019; e-mail address: Foerster_C@klinik.uni-wuerzburg.de.

In: Biochemistry Researcher Biographical Sketches ... ISBN: 978-1-62081-308-9
Editors: J. T. Walker and C. M. Jones

Chapter 134

Glucocorticoid Therapy for Cochlear Injuries: Pharmacology and Evidence from Animal Experiments

Keiji Tabuchi [*]***, Keiko Oikawa, Bungo Nishimura, Yuki Hirose, Shuho Tanaka and Akira Hara***
Department of Otolaryngology, Graduate School of Comprehensive Human Sciences, University of Tsukuba, Tsukuba, Japan

Research Summary

Dysfunction of the cochlea causes sensorineural hearing loss. Glucocorticoids show protective effects against various cochlear injuries and have been clinically applied for idiopathic sudden sensorineural hearing loss, acoustic injury, Meniere's disease, and immune-mediated hearing loss for decades. This article was aimed to give an overview of the involvement of glucocorticoids in the pathophysiology of the inner ear. Especially, recent findings derived from animal experiments were reviewed. Glucocorticoids can penetrate the blood-cochlear barrier, although they cannot permeate the normal blood-brain barrier. Elimination patterns of glucocorticoids in the cochlea were found to be entirely different from the serum and hepatic tissues. Namely, serum and hepatic concentrations of glucocorticoids rapidly declined after their intravenous administration, whereas the glucocorticoid level in the cochlea gradually decreased following administration. Glucocorticoids ameliorated the cochlear ischemia-reperfusion and acoustic injuries at a relatively wide range of doses, but their therapeutic time window is considered to be short after the onset of injuries. The therapeutic actions of glucocorticoids in cochlear injuries were considered to be mediated via both genomic and non-genomic pathways, and cochlear injuries possibly alter the levels of glucocorticoid receptors in the cochlea. It is necessary to elucidate the precise mechanisms of glucocorticoids for the treatment of sensorineural hearing loss.

[*] Corresponding author: Department of Otolaryngology, Graduate School of Comprehensive Human Sciences, University of Tsukuba, 1-1-1 Tennodai, Tsukuba 305-8575, Japan, TEL & FAX: +81-298-53-3147, E-mail: ktabuchi@md.tsukuba.ac.jp.

In: Biochemistry Researcher Biographical Sketches ... ISBN: 978-1-62081-308-9
Editors: J. T. Walker and C. M. Jones

Chapter 135

Expert Commentary: Glucocorticoids and Exercise Performance in Humans: Current Status

***K. Collomp*[1,2] *and J. De Ceaurriz*[2]**
[1]Laboratoire AMAPP, Université d'Orléans
[2]Département des Analyses, Agence Française de Lutte contre le Dopage, Chatenay-Malabry, France

Research Summary

Due to fears of doping effects, the World Anti-Doping Agency (WADA) has put certain restrictions on the use of the pharmacological class of glucocorticoids (GC). Since January 2010, all GC administered by oral, intravenous, intramuscular or rectal routes have been prohibited in-competition, although not out-of-competition. In accordance with the international standard for therapeutic use exemptions, a declaration of use must be completed by the athlete for GC administered by intraarticular, periarticular, peritendinous, epidural, intradermal and inhalation routes, whereas topical preparations for auricular, buccal, dermatological, gingival, nasal, ophthalmic or perianal disorders are not prohibited and require neither a therapeutic use exemption nor a declaration of use. Curiously, a review of the current literature to determine whether local or systemic and acute or short-term administration of GC enhances human performance reveals very few papers on this topic.

To our knowledge, only one study (Kuipers et al., 2008) was conducted following local GC administration, i.e., inhalation. In this study, 28 well-trained cyclists and rowers performed three incremental cycle ergometer tests until exhaustion to measure maximal power output before and after two and four weeks of placebo or daily budesonide inhalation (800 microg). No significant difference in maximal power output was found between the placebo (376 ± 25 W) and corticosteroid (375 ± 36 W) groups during the preintervention test, and there were no significant changes in either group after the two and four weeks of intervention. From these data, it appears that short-term inhalation of GC does not improve performance, probably because of a lack of significant systemic bioavailability with such low-dose local administration. Another possibility is that a maximal test may be inappropriate to reveal the ergogenic effects of this pharmacologic class.

In: Biochemistry Researcher Biographical Sketches ...
Editors: J. T. Walker and C. M. Jones

ISBN: 978-1-62081-308-9

Chapter 136

Glucocorticoids, Thymus Function and Sex Hormone in Human Body Growing

Gordana Bjelakovic*[1]*, Ivana Stojanovic*[1]*,
***Tatjana Jevtovic-Stoimenov*[1] *and Bojko Bjelakovic*[2]**
[1]Institute of Biochemistry, University of Nis, Serbia
[2]Pediatric Clinic, Faculty of Medicine, University of Nis, Serbia

Research Summary

Glucocortioid Effects and Thymic Functions

Glucocorticoids (GCs) are pleiotropic hormones involved in many metabolic and physiologic events essential for biological functions of the human body.GCs exert the profound anti-inflammatory and immunosuppressive activity .The thymus is a central lymphoid organ in development and maintance of immunological competence. The thymus generates large numbers of functionally mature antigen specific T cells, which then migrate to the peripheral lymphoid tissue to mediate protection against different infective agents.

The thymus enlarges until late puberty and than involves involutes, the lymphocytes and epithelial cells being nearly completely replaced with fat. Administration of GCs causes thymic involution and hormonal thymectomy. Secretion of thymulin follows a circadian rhythm, peaking during the night, just opposite to circadian rhythm of GCs.

GCs may be associated with a topical and systemic adverse effects caused by prolonged or chronic activation of the hypothalamus-pituitary-adrenal (HPA) axis or by continuous high-dose treatment.

Interaction of Glucocorticoids and Sex Hormones

GCs participate in ovarian and testicular functions. GCs suppress both gonadotropin-releasing hormone (GnRH) and gonadotropins secretion inhibiting testosterone and estradiol production by testis and ovary, respectively. Psychological stress can lead to reproductive suppression. In males, decreased serum testosterone (T) is one of the first signs of stress. During the excessive exposure to GCs apoptosis is initiated in Leydig cells, the preeminent source of testosterone. Humans exposed to prolonged stress respond with the inhibition of luteinising hormone (LH) release and blocked ovulation.

Thymus is the Primary Sex Hormone-Responsive Organ

The fact that involution of thymus starts at about the time of puberty may be related to changes in the sex hormone production. Several studies suggest that thymus is the primary sex hormone-responsive organ. The thymus is the major target of estrogens in the immune system. Ovariectomy reduces serum estrogen levels and results in enlargement of the female thymus. During the pregnancy, the period in which serum estrogen levels are high, the thymus shrinks dramatically. Thymus weight and cellular composition are very sensitive to changes in androgen status. The thymus gland of male animals is enlarged under conditions of androgen deficiency.

Metabolic correlations of glucocorticoids and gonadal steroids with thymic function are poorly understood. The objective of this study was to point out the need for better understanding of the interaction of GCs therapy and sex hormones function during, first of all, period of childhood.

In: Biochemistry Researcher Biographical Sketches ... ISBN: 978-1-62081-308-9
Editors: J. T. Walker and C. M. Jones

Chapter 137

Nutritional Factors and Osteoporosis Prevention

Masayoshi Yamaguchi*
Division of Endocrinology and Metabolism and Lipids, Department of Medicine, Emory University School of Medicine, Atlanta, Georgia, US

Research Summary

Aging induces a decrease in bone mass, and osteoporosis with its accompanying decrease in bone mass is widely recognized as a major public heath problem. Bone loss with increasing age may be due to decreased osteoblastic bone formation and increased osteoclastic bone resorption. Pharmacologic and nutritional factors may prevent bone loss with aging.

Chemical compounds in food and plants, which regulate on bone metabolism, has been to be worthy of notice in the prevention of osteoporsis. Recently, it has been shown that zinc, an essential trace element, isoflavonees (including genistein and daizein), which are contained in soybeans, menaquinone-7, an analogue of vitamin K_2 which is abundant in fermented soybeans, carotenoids β-cryptoxanthin, which is abundant in *Satuma mandarin*, and *p*-hydroxycinnamic acid, a plant component, have a stimulatory effect on osteoblastic bone formation and an inhibitory effect on osteoclastic bone resorption *in vitro*, thereby increasing bone mass *in vivo*. These factors have an effect on gene expression of proteins that are related to bone formation in osteoblastic cells and bone resorption in osteoclastic cells.

Among many food and plants, moreover, the extract components obtained from wasabi leafstalk, marine alga *S. hornerli* and bee pollen products have been found to have a unique-anabolic effect on bone mass due to stimulating osteoblastic bone formation and inhibiting osteoclastic bone resorption, although the identification of active components is remains to be elucidated.

The oral administration of those factors has been shown to prevent on bone loss in ovariectomized rats, an animal model for osteoporosis, indicating a role in prevention of

* Corresponding author: E-mail: yamamasa1155@yahoo.co.jp.

osteoporosis. Dietary intake of those factors-containg foods has been shown to have a preventive effect on osteoporosis in human subjects.

Food chemical factors thus play a role in bone health and may be important in the prevention of bone loss with increasing age.

In: Biochemistry Researcher Biographical Sketches … ISBN: 978-1-62081-308-9
Editors: J. T. Walker and C. M. Jones

Chapter 138

Regulation of Phosphate Transport in Epithelia

***R. C. Khanal*[1] *and I. Nemere*[2]**
[1]Department of Food Science, University of Arkansas, Fayetteville, AR, US
[2]Department of Nutrition and Food Sciences, Utah State University, Logan, UT, US

Research Summary

Phosphate, like it's more extensively studied partner, calcium, is maintained by dietary absorption from the intestine, reabsorption/excretion by the kidney, and mobilization/ deposition in bone. Indeed in bone, it is a key signal for mineralization of the skeleton. In intestine, the steroid hormone 1,25-dihydroxyvitamin D_3 (1,25D_3) enhances intestinal phosphate transport in young, growing animals through 'pregenomic' effects, largely mediated by the 1,25D_3-MARRS (membrane associated, rapid response steroid binding) receptor and the protein kinase (PK) C signal transduction pathway. However, due to an age-related decline in PKC, hormone stimulated phosphate absorption is lost in the adult-sized animal. One of the endogenous feedback loops for intestinal phosphate absorption is provided by the metabolite 24,25-dihydroxyvitamin D_3 (24,25D_3) which is made when an animal is vitamin D sufficient. 24,25D_3 acts to decrease the specific activity of the enzyme catalase, resulting in higher levels of H_2O_2. Hydrogen peroxide in turn acts at two points: oxidation of the 1,25D_3-MARRS receptor as well as inhibition of PKC. These effects can be overcome with diets supplemented with antioxidants. Fibroblast Growth Factor (FGF)23 is another regulator of phosphate levels in the blood. FGF23 promotes excretion of phosphate and its production is enhanced by dietary phosphate loading. Conversely, FG23 knockout mice are hyperphosphatemic. In addition to these better known hormonal regulators, this chapter also considers the influence of additional phosphatonins, minhibins, estrogen, thyroid hormones, stanniocalcin, and others. Increasing our understanding of the regulation of serum phosphate levels has implications for both human health and the sustainability of agriculture.

In: Biochemistry Researcher Biographical Sketches ... ISBN: 978-1-62081-308-9
Editors: J. T. Walker and C. M. Jones

Chapter 139

Methods for the Extraction of Metabolites from Plant Tissues

A. Mohdaly*[*] *and I. Smetanska
Department of Methods in Food Biotechnology, Berlin University of Technology, Berlin, Germany

Research Summary

Foods are commonplace and much taken for granted; however, their composition and structure are by no means simple. Extraction of one or more components from a complex mixture is a requirement for many operations in the food engineering and biotechnology industries.

This technology is used for either to recover important food components, being a main processing stage for the production of certain food products (sugar, oils, proteins), to isolate desired components (antioxidants, flavors), or to remove contaminants and other undesirable components (alkaloids, cholesterol).

Extraction methods rely on exploiting differences in physical or chemical properties of the mixture of components. Some of the more common properties involved in separation processes are particle or molecular size and shape, density, solubility and electrostatic charge. In some operations, more than one of these properties are involved. However, most of the processes involved are of a physical nature. The recent progress in extraction technology has been in the use of novel techniques for separation.

In this chapter, we review the extraction methods as conventional solvent extraction, supercritical fluid extraction (SFE), aqueous two-phase extraction (ATPE), liquid membrane extraction, and reverse micelle extraction. They are applied in food processing from fundamental theory to optimum practical application through using the relevant equipment, solvents, and the appropriate methods of process optimization.

[*] Corresponding author: Adel Abdelrazek Abdelazim Mohdaly, Methods of Food Biotechnology, Berlin University of Technology, Königin-Luise-Str.22 D-14195 Berlin, Germany, E-mail: adola2010@yahoo.com;
Food Science and Technology, Faculty of Agriculture, Fayoum University, Egypt.

In: Biochemistry Researcher Biographical Sketches … ISBN: 978-1-62081-308-9
Editors: J. T. Walker and C. M. Jones

Chapter 140

Application of 'Omics' Technologies for Improvement of Meat Quality

***Urmila Basu*[1*], *Le Luo Guan*[1,*,#], *Masaaki Taniguchi*[2], *Yongmei Zhao*[1,3] *and Michael V. Dodson*[4]**
[1]Department of Agricultural, Food and Nutritional Science,
University of Alberta, Edmonton, Alberta, Canada
[2]National Institute of Agrobiological Sciences, Tsukuba, Ibaraki, Japan
[3]Department of Life Science, Xi'an University of Arts and Science, Shaanxi, P. R. China
[4]Department of Animal Science, Washington State University, Pullman, WA, US

Research Summary

Meat quality is the key to both industry and consumers. Major aspects of quality of the meat (tenderness, flavor, and nutritional value) are being studied via use of molecular biology methods. Indeed, the recent technological innovations in functional genomics/proteomics including the global analysis of RNA, protein expression, protein localization and protein-protein interactions may provide new insights to understand the molecular basis of meat quality. Moreover, interpreting the complex interplay of gene expression events involved in the process of carcass formation will enhance our understanding of metabolism and optimizing animal nutritional status in the 21st Century. This review aims to highlight the recent achievements on our understanding of the complexity of the molecular processes and their interaction with environmental factors associated with meat quality, using different tools of genomics, transcriptomics and proteomics. Furthermore, combining a holistic view of phenotypes at the molecular level with genetic and protein markers seem to be a particularly promising approach for improving meat quality in livestock species. Collectively, knowledge gained from these approaches may be applied in defining and improving management systems for the nutritional component and quality of meat for market needs.

[#] Corresponding author: Tel: +1 780 492 2480. Fax: +1 780 492 4265. E-mail: lguan@ualberta.ca.
[*] These authors contrubute equally to this chapter.

In: Biochemistry Researcher Biographical Sketches … ISBN: 978-1-62081-308-9
Editors: J. T. Walker and C. M. Jones

Chapter 141

Role of Chaperones in Dystrophic and Senescent Skeletal Muscle Fibres

Pamela Donoghue, Philip Doran and Kay Ohlendieck
Department of Biology, National University of Ireland, Maynooth, County Kildare, Ireland

Research Summary

In skeletal muscles, stress proteins are essential for cellular survival during fibre injury. The natural response to stressful conditions in muscle involves both the class of high-molecular-mass heat shock proteins, and members of the sub-family of low-molecular-mass stress proteins that are characterized by a conserved carboxy-terminal sequence, the 90-residue α-crystallin domain. Proteomic profiling studies of dystrophic and aged skeletal muscles have identified a greatly increased expression of certain small heat shock proteins, such as αB-crystallin and cvHsp. Both small heat shock proteins are proposed to play a critical role in preventing deleterious muscle protein aggregation in the plasmalemma, cytosol and contractile apparatus. This chapter will discuss the findings from biochemical and proteomic studies into the role of chaperones during skeletal muscle aging and X-linked muscular dystrophy. Sarcopenia, the age-dependent decrease in skeletal muscle mass and contractile function, is one of the most striking features of aged mammalians. Loss of muscle strength and bone integrity is responsible for frequent falls, fractures and poor balance in many older patients. Duchenne muscular dystrophy represents the most common gender-specific genetic disease and is characterized by progressive muscle wasting due to the lack of the membrane cytoskeletal protein dystrophin. Here we discuss changes in molecular chaperones in these two major muscle diseases. That stress proteins are of critical importance for the maintenance of neuromuscular integrity has been shown by many biochemical studies. For example, the abundant muscle chaperone Hsp70 plays a key role in maintaining fibre homeostasis by stabilizing metabolic pathways, preventing excess apoptosis and facilitating muscular adaptation processes. Over-expression of this chaperone in the muscle of senescent transgenic mice was shown to protect fibres against age-related muscle dysfunction, highlighting the crucial function of stress proteins during muscle aging. In addition, experimental treatment protocols for muscular dystrophy showed that the increased expression of heat shock proteins in myoblasts improved the cellular survival rate following transplantation. Therefore, the recently established up-regulation of small heat shock proteins in aged and dystrophic muscles could be considered an auto-protective mechanism in degenerating fibres. This review critically examines whether a pharmacologically enhanced stress response in diseased skeletal muscles may be exploitable as a new approach to counteract excessive muscle damage in X-linked muscular dystrophy or sarcopenia of old age.

In: Biochemistry Researcher Biographical Sketches ... ISBN: 978-1-62081-308-9
Editors: J. T. Walker and C. M. Jones

Chapter 142

Amino Acid Composition of Some Aquaculture Fauna Resources in Nigeria

*E. I. Adeyeye**
Department of Chemistry, University of Ado Ekiti, Ado-Ekiti, Nigeria

Research Summary

Amino acid composition of three fin fishes: *Clarias anguillaris, Oreochromis niloticus* and *Cynoglossus senegalensis;* fresh water shell fish: flesh, exoskeleton and whole organism of each sex of crab (*Sudananautes africanus africanus)* and marine water shell fish: whole organism, flesh and exoskeleton of pink shrimp (*Penaeus notabilis)* was determined on dry weight basis. Glutamic acid was the highest concentrated amino acid in all the samples except the whole body of pink shrimp where aspartic acid was highest (all reported in mg/g crude protein). In the amino acid scores (based on whole hen's egg), highest score was Gly (1.04, *C. anguillaris),* His (0.89, *O. niloticus),* Glu (0.96, *C. senegalensis);* His (1.36, shrimp exoskeleton), Gly (1.61, shrimp flesh), His (1.11, shrimp whole body); Gly (1.50-1.67, female crab whole body, flesh, exoskeleton) and Gly (1.37-2.06, male crab whole body, flesh, exoskeleton). Essential amino acid highest scores were: (on whole hen's egg), Lys (0.81, *C. anguillaris),* His (0.89, *O.niloticus),* Arg (0.86, *C. senegalensis)*; His (1.11 – 1.36, shrimp samples); Lys (0.79-1.07, female crab samples); Lys (0.94, male crab whole body), His (0.84, male crab flesh), His (0.74, male crab exoskeleton) whereas the limiting essential amino acids here were Thr for all the samples. On provisional essential amino acid scoring pattern: Phe + Tyr was highest in all the samples with values ranging from 1.01-1.38 and their limiting essential amino acid scores was Thr. Essential amino acid scores based on pattern for pre-school child were highest in Met + Cys (1.17-1.34, fish); His (1.61-1.72, flesh, shell of shrimp), Ile (1.58, shrimp whole body); Ile (1.41-1.45, whole body and exoskeleton of female crab), Met + Cys (1.44, flesh of female crab); Met + Cys (1.16-1.32, male crab samples)

* E-mail: eiadeyeye@yahoo.com; Tel: +2348035782925.

whereas their limiting essential amino acids on this standard were: Thr (fish); Thr (shrimp); Lys for female and male crab samples. Predicted protein efficiency ratio (P-PER) was 1.89-2.22 (fish); 2.41-3.24 (shrimp); 2.6-3.4 (female crab); 2.4-2.9 (male crab). Essential amino acid index range were: 0.85-0.92 (fish); 1.06-1.21 (shrimp); 1.06-1.16 (female crab); 0.92-1.10 (male crab). The Leu/Ile ratios were all low with values of 2.0-2.6 (fish); 1.1-1.6 (shrimp); 1.3-1.8 (female crab) and 1.5-2.2 (male crab). The samples isoelectric points (pI) were determined for quick determination of their protein isolates. The samples were good sources of essential amino acids with no likelihood of concentration antagonism.

In: Biochemistry Researcher Biographical Sketches … ISBN: 978-1-62081-308-9
Editors: J. T. Walker and C. M. Jones

Chapter 143

The Biochemistry of Isoelectric Processing and Nutritional Quality of Proteins and Lipids Recovered with This Technique

C. K. Gehring, J. C. Gigliotti, J. C. Tou, J. S. Moritz and J. Jaczynski*
West Virginia University, Division of Animal and Nutritional Sciences, Morgantown, WV, US

Research Summary

The isoelectric behavior of food proteins has been well-characterized in the scientific literature. The isoelectric point (pI) of a protein is a pH at which the protein maintains a zero net electrostatic charge. In this state, protein-protein hydrophobic interactions overcome protein-water electrostatic interactions resulting in minimum protein solubility (i.e., isoelectric precipitation). Conversely, isoelectric solubilization occurs when the pH of the protein solution is altered away from the pI. In this state, protein-water interactions are favored in addition to protein-protein electrostatic repulsion. Therefore, protein solubility/insolubility can be induced by isoelectric solubilization/precipitation, respectively. Consequently, this biochemical concept can be used to recover nutrients such as proteins and lipids from food animal processing by-products.

Food processing inevitably generates significant amounts of by-products. The by-products contain highly nutritive proteins and lipids that, if recovered, could be diverted from disposal or low-value utilization and instead be used for human and animal consumption. The processing of aquatic animals for human food products (i.e., seafood) typically generates more by-products than the processing of farm animals. Not only is seafood processing

* Corresponding author: Tel: 304-293-2406 x 4445; Fax: 304-293-2232; E-mail: Jacek.Jaczynski@mail.wvu.edu.

inherently less efficient, but generally fish stocks are declining and many of the world fisheries are already over-exploited, which may lead to their collapse by the mid-century.

The processing by-products of seafood as well as farm animals retain ample muscle proteins and oil. The oil from aquatic animals typically contains omega-3 fatty acids; while the proteins in the processing by-products generally contain all of the essential amino acids in adequate quantity to support human and animal growth. Therefore, efficient recovery of muscle proteins and lipids from processing by-products of food animals could provide nutritious and inexpensive human food products; as well as reduce the waste or low-value utilization that currently is the norm. However, these tremendous resources are currently unavailable for human and animal consumption due to the lack of commercially available technology to recover proteins and lipids from food animal processing by-products and underutilized low-value aquatic species. Isoelectric solubilization/precipitation allows selective water-solubility of proteins. Therefore, while proteins are solubilized, all other components may be separated by for example centrifugation; followed by protein precipitation and recovery. The centrifugation also separates lipids.

Consequently, several laboratories are actively researching the application of the isoelectric point (pI) to recover functional muscle proteins and lipids followed by the determination of nutritional quality of the recovered nutrients for potential applications to human food products and animal feeds. This chapter reviews the fundamental biochemical principles of isoelectric behavior of food proteins and lipids as well as their structure and interaction with water. In addition, this chapter discusses nutritional value of the proteins and lipids recovered with isoelectric processing and their potential use in food products for human consumption as well as animal feeds.

In: Biochemistry Researcher Biographical Sketches ... ISBN: 978-1-62081-308-9
Editors: J. T. Walker and C. M. Jones

Chapter 144

Nutritional Biochemistry of Curcumin (Diferuloylmethane) and a Review of Its Biological Actions on Articular Chondrocytes

Ali Mobasheri[1,,#], Yves Henrotin[2#], Abigail L. Clutterbuck[1], David Allaway[3], Emma M. Lodwig[3], Pat Harris[3], Marianne Mathy-Hartert[2] and Mehdi Shakibaei[4]*

[1]Division of Veterinary Medicine, School of Veterinary Medicine and Science, University of Nottingham, Sutton Bonington Campus, Sutton Bonington, Leicestershire, UK

[2]Bone and Cartilage Research Unit, University of Liège, Institute of Pathology, Sart-Tilman, Belgium

[3]WALTHAM Centre for Pet Nutrition, Waltham-on-the-Wolds, Melton Mowbray, Leicestershire, UK

[4]Musculoskeletal Research Group, Institute of Anatomy, Ludwig-Maximilians-University Munich, Munich, Germany

Research Summary

Curcumin (diferuloylmethane) is the principle biochemical component of the spice turmeric and has been shown to possess potent anti-catabolic, anti-inflammatory and anti-oxidant, properties. In this chapter we provide a brief review of osteoarthritis (OA) and discuss the anti-inflammatory properties of curcumin in the context of its actions on articular chondrocytes. We highlight its potential benefits and drawbacks for treating OA and related osteoarticular conditions. Recent work has shown that curcumin protects human chondrocytes

* Corresponding author: E-mail: ali.mobasheri@nottingham.ac.uk, Phone: +44-1159516449; Fax; +44-1159516440.
# These authors contributed equally to this article.

from the catabolic actions of interleukin (IL)-1 beta (IL-1β) including matrix metalloproteinase-3 (MMP-3) upregulation, inhibition of collagen type II and down-regulation of β1-integrin expression. Curcumin blocks IL-1β-induced proteoglycan degradation, Nuclear Factor Kappa B (NF-κB) signaling, chondrocyte apoptosis and activation of caspase-3. The available data from *in vitro* and *in vivo* studies suggest that curcumin may be a beneficial complementary treatment for OA in humans and companion animals. Nevertheless, before initiating extensive clinical trials, more basic research is required to improve its absorption and bioavailability and gain additional information about its safety and efficacy in different species. Once these obstacles have been overcome, curcumin and structurally related biochemicals could become suitable nutritional and nutraceutical alternatives to the non-steroidal anti-inflammatory drugs that are currently used for the treatment of OA.

In: Biochemistry Researcher Biographical Sketches ... ISBN: 978-1-62081-308-9
Editors: J. T. Walker and C. M. Jones

Chapter 145

Effect of Creatine Applied as Food Supplement on Human Metabolism

***Tomáš Navrátil*[1,2*]*, Eva Kohlíková*[3]*, Miroslav Petr*[4]
and Kamila Přistoupilová[5]**
[1]J. Heyrovský Institute of Physical Chemistry of the ASCR, v.v.i.,
Department of Biophysical Chemistry, Prague, Czech Republic
[2]Charles University in Prague, 1st Faculty of Medicine,
Institute of Medical Biochemistry, Prague, Czech Republic
[3]Department of Physiology and Biochemistry, Faculty of Physical Education and Sport,
Charles University in Prague, José Martího, Prague, Czech Republic
[4]GHC Genetics, Prague, Czech Republic
[5]Prague, Czech Republic

Research Summary

Creatine (CR) (methyl guanidine acetic acid), is recommended for application as food (dietary) supplement. It is supposed that human organism uses it for formation of CR phosphate, which is necessary as a source of energy for muscular work. The fate of supplemented CR is not as simple as that. Its molecule is decomposed and reused again in each recipient individually.

In our short-term study we have observed changes in the metabolism of two-carbon (2C) units and of thiolic substances 4 to 6 hours after CR supplementation. The level of thiodiglycolic acid (TDGA) in urine increased rapidly and then decreased again to the original level. At the same time the pH values of urine increased by 1.5 units. The level of TDGA in urine is a marker of disturbance in metabolic pathways of 2C units and of thiolic compounds.

Our long-term CR study, with quantitative evaluation of more than 30 parameters of potential metabolic products revealed that the biochemical pathways, affected by CR

* Corresponding author: E-mail: navratil@jh-inst.cas.cz.

application included intrinsic participation of vitamin B_{12} and folates. The levels of those two vitamins changed in a mutually reverse way. The men under study can be divided into 4 different groups according to changes of CR levels found in urine, and of levels of folates and vitamin B_{12}, determined in blood before and after CR administration. The probands of each group utilized CR as donor of 1C and 2C units derived from CR in different metabolic pathways. Differences between initial and terminal TDGA levels indicated that CR disturbed the equilibria of redox processes, catalyzed by folates and vitamin B_{12}. The levels of homocysteine increased in all but one probands. The amount of creatinine excreted into urine depended on the extent of metabolic disturbances. The observed changes indicated that exogenous application of CR affected metabolic pathways connected with endogenous synthesis as well as with decomposition of CR.

In: Biochemistry Researcher Biographical Sketches … ISBN: 978-1-62081-308-9
Editors: J. T. Walker and C. M. Jones

Chapter 146

Production, Properties and Stability of Chicken Meat Protein Hydrolysate Powder

Louise Emy Kurozawa[1], Kil Jin Park[2] and Miriam Dupas Hubinger[1]
[1]Department of Food Engineering, Faculty of Food Engineering, University of Campinas, SP, Brazil
[2]Faculty of Agricultural Engineering, University of Campinas, SP, Brazil

Research Summary

Protein hydrolysates are mainly applied in the nutritional management of individuals who cannot digest whole/intact protein. The findings of numerous studies have confirmed that protein hydrolysate has many dietary uses due to its high nutritional and therapeutic values, such as reduced immunological reactivities, high biological value peptides and antioxidant activity. They can be obtained by acid/basic or enzymatic hydrolysis. In enzymatic reactions, it is important to know which variables, such as temperature, pH, enzyme concentration and substrate:water ratio are critical, since the optimization of process variables is essential to obtain an adequate product. The present chapter focuses on the several aspects of enzymatic hydrolysis of meat protein, such as reaction process and nutritional and functional properties of protein hydrolysate. Moreover, due to the fact that there are scarce results for chicken breast meat, the main information to obtain protein hydrolysate from this sort of muscle is shown in this chapter: enzymatic reaction parameters, kinetics of the process, protein recovery and amino acid composition of protein hydrolysate. As protein hydrolysates are highly perishable due to their high moisture and protein content, they have to pass through an additional process in order to enlarge their shelf life. Spray drying can be used and involves both particle formation and drying. The drawback is that spray drying of protein hydrolysates is difficult, mainly due to the low molecular weight peptides, presenting low glass transition temperature (Tg) values and, consequently, high hygroscopicity and thermoplasticity. Since the Tg increases with molecular weight, the addition of maltodextrin and gum Arabic has been used in the production of the chicken meat protein hydrolysate powder, reducing the stickiness and wall deposition in spray drying and rising powder stability storage. Therefore, the spray drying process of chicken breast meat hydrolysate and its powder stability are described and discussed in this chapter.

In: Biochemistry Researcher Biographical Sketches ...
Editors: J. T. Walker and C. M. Jones
ISBN: 978-1-62081-308-9

Chapter 147

Alpha-Galactosidase: A Food and Feed Enzyme

G. S. Anisha[1], Rojan P. John[2] and Ashok Pandey[3]
[1]Department of Zoology, Government College, Chittur, Palakkad, India
[2]Institut national de la recherche scientifique-Eau Terre Environnement, Québec (QC), Canada
[3]Biotechnology Division, National Institute for Interdisciplinary Science and Technology, CSIR, Trivandrum, India

Research Summary

Legumes have traditionally been an important part of the diets of many cultures throughout the world. The nutritional profile of beans shows that they have much to offer; beans are high in protein, low in saturated fat, and high in complex carbohydrates and fiber. Beans are also a good source of several micronutrients and phytochemicals. Despite the nutritional and health benefits of legumes their utilization is constrained due to the presence of raffinose family oligosaccharides which have the tendency to induce flatulence. It is desirable that these flatulence factors are removed from the legumes, if they are to be widely accepted for nutritional utilization. Various conventional methods like soaking, cooking, germination etc. are in practice for the removal of these anti-nutritive flatulence-causing factors from the legumes. But these methods have several drawbacks and they affect the physicochemical characteristics of legumes. Of the various techniques used, enzymatic treatment of legumes has been found most effective. Alpha-galactosidase is a versatile enzyme which can alleviate the problems of flatulence by mediating the removal of flatulence factors from legumes. It is an exoglycosidase which can catalyze the hydrolysis of raffinose oligosaccharides. Various fermentation technologies like submerged and solid-state fermentation are practiced for the microbial production of alpha-galactosidase. Several reports are available in the literature documenting the use of microbial alpha-galactosidase for the treatment of legumes to be used for human consumption as well as feed for animals.

Usually alpha-galactosidases for legume processing should be stable at high temperatures and over a wide range of pH. Microbial alpha-galactosidases are also commercially available as enzyme supplements in dietary enzyme products.

In: Biochemistry Researcher Biographical Sketches ... ISBN: 978-1-62081-308-9
Editors: J. T. Walker and C. M. Jones

Chapter 148

Foods of Plant Origin as Source of Nitric Oxide Production Inhibitors

Filomena Conforti* and Francesco Menichini
University of Calabria, Department of Pharmaceutical Sciences, Italy

Research Summary

Nitric oxide (NO) is a diatomic free radical produced from L-arginine by constitutive and inducible nitric oxide synthase (cNOS and iNOS) in numerous mammalian cells and tissues. Nitric oxide (NO), superoxide (O^{2-}) and their reaction product peroxynitrite ($ONOO^{-}$) may be generated in excess during the host response against viral and antibacterial infections and contribute to some pathogenesis by promoting oxidative stress, tissue injury and, even, cancer. Oxidative damage, caused by action of free radicals, may initiate and promote the progression of a number of chronic diseases, including cancer, cardiovascular diseases, Alzheimer's disease, diabetes and inflammation. The mechanism of inflammation injury is attributed, in part, to release of reactive oxygen species from activated neutrophils and macrophages. ROS propagate inflammation by stimulating release of mediators such as NO and cytokines. The interest of the research is motivated by the current need to find new substances of natural origin which have demonstrated effectiveness in the described fields of application and low degree of toxicity for humans. Natural products provide a vast pool of NO inhibitors that can possibly be developed into clinical products. This article reviews various extracts and secondary metabolites from foods of plant origin with NO inhibitory activity that can be focused for drug development programs.

* Corresponding author: Department of Pharmaceutical Sciences, University of Calabria, I-87036 Rende (CS), Italy, Tel: +39 0984 493168, Fax: +39 0984 493298, E-mail: filomena.conforti@unical.it.

In: Biochemistry Researcher Biographical Sketches ...
Editors: J. T. Walker and C. M. Jones

ISBN: 978-1-62081-308-9

Chapter 149

Application of Plant-Derived Food Lectins in Proteoglycomics and Immunomodulation

Marija Gavrovic-Jankulovic* and Milica Grozdanovic
Faculty of Chemistry, Department of Biochemistry, Belgrade, Serbia

Research Summary

Lectins are carbohydrate-binding proteins that discriminate sugars based on subtle difference in structure. The biological role of intracellular lectins has not been completely elucidated yet; however, it is known that lectins exposed on the cell membrane are involved in various biological processes: cell-to-cell communication, cell migration, and the lectin pathway of complement activation. The functional significance of carbohydrates, particularly those attached to proteins, has been determined in health investigations and imposed structural investigations on lectins, as well as by their application in research and medicine. Considerable knowledge gained on the structure and carbohydrate specificity of plant lectins established them as a good tool for profiling carbohydrate structures present on cell-surfaces. A lectin microarray approach rapidly evaluates the carbohydrate composition on the surface of intact cells, and provides means to elucidate the dynamic role of sugars in biological systems. The importance of protein glycosylation has been implicated in designing novel therapeutic approaches in certain diseases. Due to their unique carbohydrate specificity, besides being applicable in deciphering the structural diversity of carbohydrates, plant lectins exhibit modulatory potential on the immune system. Certain plant lectins are prone to induce immune response upon oral feeding. Therefore, a search for novel immunostimulatory molecules and adjuvants, in order to enhance or to direct an appropriate immune response against target immunogens, has been a major and recurrent issue in vaccine development.

* Corresponding author: mgavrov@chem.bg.ac.rs, Tel: +381 11 3336 661, Fax: +381 11 2184 330.

In: Biochemistry Researcher Biographical Sketches ... ISBN: 978-1-62081-308-9
Editors: J. T. Walker and C. M. Jones

Chapter 150

Biochemical and Proteomic Profiling of Key Metabolic Enzymes in Aging Skeletal Muscle

Pamela Donoghue, Philip Doran and Kay Ohlendieck
Department of Biology, National University of Ireland, Maynooth, County Kildare, Ireland

Research Summary

The age-dependent decrease in skeletal muscle mass and function represents a serious pathobiochemical and metabolic problem for the senescent body. The metabolic alterations associated with fibre type shifting and the decline in contractile strength in senescent fibres, now generally referred to as sarcopenia of old age, plays a central role in the physical weakness of many elderly patients. Nutritional supplementation regimes, high-protein diets and the abolishment of a sedentary lifestyle have been shown to partially reverse the symptoms of age-related muscle wasting. However, no nutritional or medical treatment option is currently available that can entirely halt or reverse fibre degeneration in the senescent organism. It is hoped that the future elucidation of the molecular and cellular mechanisms underlying fibre aging will lead to the design of superior treatment strategies to counteract sarcopenia-associated fragility. The mass spectrometry-based proteomic profiling of key metabolic pathways suggests itself as a highly efficient analytical tool that could establish a biomarker signature for sarcopenia. The identification of new therapeutic targets may be useful for the development of advanced diagnostic procedures and novel pharmacological options. This chapter outlines the recent proteomic identification of metabolic biomarkers of sarcopenia and discusses how a shift to more aerobic-oxidative metabolism in a slower-twitching fibre population affects the overall physiology of aged muscles. This includes the detailed analysis of the expression and post-translational changes in pyruvate kinase and related glycolytic enzymes, the proteomic profiling of key mitochondrial components and the biochemical analysis of metabolic transporters. Since adult skeletal muscles naturally exhibit a dynamic fibre type distribution pattern and show a high degree of intrinsic tissue plasticity, biochemical findings from denervation experiments, surveys of exercise programs, fibre typing and fast-to-slow transformation studies are compared to the age-dependent changes in metabolic muscle markers. In the long-term, the establishment of a comprehensive biomarker signature of metabolic alterations in aged fibres may lead to the design of improved nutritional supplementation programs, better support for the senescent metabolism, optimized resistance training regimes and the prevention of obesity-related complications in the aging population.

In: Biochemistry Researcher Biographical Sketches ... ISBN: 978-1-62081-308-9
Editors: J. T. Walker and C. M. Jones

Chapter 151

Inhibitory Mechanism of Longer Chain Fatty Acids on Mammalian DNA Polymerase β Activity

Yoshiyuki Mizushina[1,2,*]***, Kengo Sakaguchi***[3]***,***
Fumio Sugawara[3] ***and Hiromi Yoshida***[1,2]
[1] Laboratory of Food & Nutritional Sciences,
Department of Nutritional Science, Kobe-Gakuin University,
Nishi-ku, Kobe, Hyogo, Japan
[2] Cooperative Research Center of Life Sciences,
Kobe-Gakuin University, Chuo-ku, Kobe, Hyogo, Japan
[3] Department of Applied Biological Science, Faculty of Science and Technology,
Tokyo University of Science, Noda, Chiba, Japan

Research Summary

Fatty acids, which are major food components and nutrients, can inhibit mammalian DNA polymerase (pol) activities *in vitro*, and the mode of this inhibition is reviewed. The essential moieties of linear longer chain fatty acids for pol inhibition are below: 1) C18- or more carbon chain, 2) a free carboxylic group, and 3) double bonds of *cis*-configuration. Among the C18- to C24-fatty acids examined, the strongest inhibitor was a C24-fatty acid, nervonic acid (NA). For pol β, this is a repair-related pol, NA acted by competing with both the nucleotide substrate and DNA template-primer, but for pol α, which is a replicative pol, NA acted non-competitively. NA-binding to pol β could be stopped with a non-ionic detergent but the binding to pol α could not. Pol β (39 kDa enzyme) consists of 2 distinct domains: the N-terminal "DNA template binding" domain (8 kDa) and the C-terminal "catalytic" domain (31 kDa). NA was found to bind to the 8 kDa domain fragment but did not

[*] Corresponding author: Yoshiyuki Mizushina Ph.D., Tel: +81-78-974-1551 (ext.3232); Fax: +81-78-974-5689; E-mail: mizushin@nutr.kobegakuin.ac.jp.

interact with the 31 kDa domain. Structural interaction with the 8 kDa domain of pol β and NA was analyzed by two-dimensional NMR (^{1}H - ^{15}N HMQC). In the 8 kDa domain with NA, the structure that forms the interaction interface included helix-1, helix-2, helix-4, the three turns (residues 1–13, 48–51 and 79–87) and residues adjacent to an Ω-type loop connecting helix-1 and helix-2 of the same face. No significant shifts were observed for any of the residues on the opposite side of the 8 kDa domain. The three-dimensional docking simulation was modeled so that the 8 kDa domain of pol β bound to NA as a 1:1 complex. The relationship between the structure of fatty acids and the inhibition of mammalian pols, especially pol β, is discussed.

In: Biochemistry Researcher Biographical Sketches ... ISBN: 978-1-62081-308-9
Editors: J. T. Walker and C. M. Jones

Chapter 152

Endorhizal Fungi Associated with Vascular Plants on Truelove Lowland, Devon Island, Nunavut, Canadian High Arctic

Catherine Peters[1], James F. Basinger[2] and Susan G. W. Kaminskyj[1*]
[1]Department of Biology, University of Saskatchewan, Canada
[2]Department of Geological Sciences, University of Saskatchewan, Canada

Research Summary

Truelove Lowland on Devon Island, Nunavut (74 °N) in the eastern Canadian High Arctic Archipelago, has long been investigated for its flora, fauna, and microbiota. Nevertheless, apart from ectomycorrhizae associated with *Salix* and *Dryas*, the record of fungal-plant root interactions has been described as sparse or absent. In the present study, 20 samples of roots and associated soils (55 plants in total) from ten genera in nine families were collected during the summer of 2006. These were analyzed for endorhizal fungi using our recently developed high-sensitivity lactofuchsin epifluorescence method. Endorhizal colonization was assessed using a quantitative microintersect method that distinguishes among fungal morphotypes.

Of the total 958 intersections assessed at 400 X total magnification for endorhizal colonization, only 154 lacked fungi, and many intersections had more than one endorhizal morphotype. Most colonization was by septate endophytes (average abundance 66 %, range 13 - 100 %), and fine endophytes (average abundance 48 %, range 0 - 100 %). Endorhizal morphology in *Dryas* and *Saxifraga* roots was superficially similar to both ericoid mycorrhizae and ectendomycorrhizae: an extraradical fungal sheath that also grew between and within root cortical cells. Those definitions had been restricted to the Ericaceae and

[*] Corresponding author: Tel 306-966-4422; Email susan.kaminskyj@usask.ca

Pinaceae, respectively. We propose a modified description, ectendiform, for this endorhizal morphotype, without restricting usage to specific plant families.

Soil samples in which the Truelove plants had grown, which had been stored at -20 °C, were planted with wheat in July 2007 and grown in a greenhouse for 10 weeks. Fungal colonization abundance in these wheat roots ranged from 35 % to 100 %. Our results show that fine endophyte and septate endophyte fungi are prevalent in the fungal endorhizal diversity of the Truelove Lowland tundra ecosystem, and are predicted to have a correspondingly important role in plant adaptation to tundra environments.

In: Biochemistry Researcher Biographical Sketches ... ISBN: 978-1-62081-308-9
Editors: J. T. Walker and C. M. Jones

Chapter 153

Involvement of the Mitochondrial ATP-Sensitive Potassium Channel in the Beneficial Effects of Fasting on the Ischemic-Reperfused Rat Heart

M. G. Marina Prendes, M. S. González, R. Hermann, N. G. Pascale, M. E. Torresín, M. M. Jaitovich, E. A. Savino and A. Varela
Department of Biological Sciences, Faculty of Pharmacy and Biochemistry,
University of Buenos Aires and IQUIMEFA-CONICET, Buenos Aires, Argentina

Research Summary

1. Fasting improves contractile recovery and attenuates lactate production and mitochondrial permeability transition (MPT) without altering cell viability in ischemic-reperfused rat hearts. 5-hydroxydecanoate (5-HD), a mitochondrial ATP-sensitive K^+ channel ($K\text{-}ATP_{mito}$) blocker, abolishes the improvement of mechanical recovery elicited by fasting despite it decreases lactate production, suggesting that $K\text{-}ATP_{mito}$ might be involved in the preservation of contractility. This chapter assessed the contribution of $K\text{-}ATP_{mito}$ in the attenuation of MPT which in turn may improve mitochondrial energetics.

2. Langendorff-perfused hearts from fed and fasted rats were subjected to ischemia-reperfusion in the presence or absence of 5-HD. To assess whether 5-HD has any direct effect on glycolysis, a cell free heart extract containing all the glycolytic enzymes was used. MPT was quantitated measuring the mitochondrial 2-[^{3}H]-deoxyglucose (^{3}H-2-DG) entrapment. Total heart ^{3}H-2-DG content as an estimation of necrosis was measured. As an index of mitochondrial energetic function, the rate of ATP synthesis was examined in mitochondria isolated from the ischemic-reperfused hearts.

3. Fasting increased the rate of mitochondrial ATP synthesis and attenuated MPT. 5-HD abolished these protective effects. 5-HD did not change total heart ^{3}H-2-DG content indicating that it lacked effects on cell survival. Since 5-HD did not affect glucose

consumption and lactate production in the cell-free heart-extract the inhibition of glycolysis could be due to its own oxidative metabolism.

4. It may be concluded that cardioprotection elicited by fasting in the ischemic-reperfused heart could be ascribed, at least in part, to the K-ATP$_{mito}$ activation which in turn inhibits MPT and induces energetic preservation.

In: Biochemistry Researcher Biographical Sketches ... ISBN: 978-1-62081-308-9
Editors: J. T. Walker and C. M. Jones

Chapter 154

Application of Artificial Neural Network for Predicting Physicochemical Properties of Pressure-Induced Gel from Industrial Milk Whey Proteins

***Tai-Hua Mu*[1*] *and Jin-Song He*[2]**
[1]Institute of Agro-Food Science & Technology,
Chinese Academy of Agricultural Sciences, Beijing, China
[2]Department of Environmental Science and Engineering, Tsinghua University,
Beijing, China

Research Summary

The effects of high hydrostatic pressure, protein concentration, and sugar concentration on the gelation of a whey protein isolate (WPI) were investigated. The WPI solutions with different concentrations in the presence or absence of lactose (0-20% w/v) were pressurized at 200-1000 MPa and 30°C for 10 min. The hardness and breaking stress of pressure-induced WPI gels increased with increasing concentration of WPI (12-18%) and hydrostatic pressure. Lactose decreased the hardness and breaking stress of the WPI gel. Furthermore, these results were used to establish the artificial neural network (ANN).

Since physical property changes are important gel attributes applied in food manufacture, the ability to predict the physical properties of gel is of great interest. One objective of this research was to predict the physical properties of the pressure-induced gel via the use of an ANN. A multiple layer feed-forward neural network was also established to predict the physical properties including the hardness and breaking stress of the gel based on the inputs of the pressure, protein concentration, and sugar concentration.

* Corresponding author: Tel./Fax:+86-10-6281-5541, E-mail: mutaihua@126.com.

ANN models were then tested against an independent dataset. A useful prediction was possible, as measured by a low mean square error (MSE < 0.05) and a regression coefficient ($R^2 > 0.99$) between true and predicted data in all cases.

In: Biochemistry Researcher Biographical Sketches ...
Editors: J. T. Walker and C. M. Jones

ISBN: 978-1-62081-308-9

Chapter 155

Isoflavones: a Race after the Rescue of the Ageing Hippocampus

***Khalid Jamali*[1,2], *Aline Marighetto*[3] and *Catherine Bennetau-Pelissero*[1,2*]**

[1]Université de Bordeaux, UMRS, ENITA de Bordeaux, France
[2]Université de Bordeaux, Bordeaux, France
[3]Université de Bordeaux, CNIC, Talence, France

Research Summary

Isoflavones are naturally occurring molecules present in significant amount in many leguminous, some of which are edible plants. Isoflavones exhibit similarities in their chemical structure with steroidal estrogens. This allows them to trigger some of the endogenous physiocellular signalization processes of estrogens that are known to occur naturally within living organisms. In fact, isoflavones mediate numerous health effects mainly related to their estrogenic and/or anti-estrogenic actions. Some of these effects are estrogen receptor-mediated physiological phenomenona such as those linked to the hypothalamo-hypophyso-gonadal axis. Moreover, because the expression of estrogen receptors is highly enriched in neural tissues, it was not surprising to found out that isoflavones were potent neurotrophic and neuroprotective compounds as it is the case for estrogens. Functional estrogen receptor proteins are also expressed in some central nervous system territories, including the hippocampus and the cerebral cortex, that are well thought to be devoted to higher cognitive brain functions such as the memory function. More particularly, the hippocampus expresses both nuclear and membrane estrogen receptors known to mediate local physiocellular alterations of the neuronal phenotype, namely dendritic spine development, synaptic plasticity and modulation of the Ca^{++} trafficking rate between different compartments of the cell. These physiocellular effects are thought to be part of the adaptative chemical and architectural

* Corresponding author: Unité Micronutriments Reproduction Santé, ENITA de Bordeaux, 1, cours du Général de Gaulle, 33 175 Gradignan FRANCE, Tel : 33+(0)5 57 35 07 52, Fax : 33+(0)5 57 35 07 39, e-mail : c-bennetau@enitab.fr

changes sustaining some processes of the mnemonic information storage. This review provides *in vitro*, animal, and clinical data intending to highlight the impact of isoflavones on the physiocellular features of the hippocampal formation plasticity and, therefore, on the hippocampus-dependent cognitive brain function. In summary, isoflavones not always exhibit all the mechanistic effects of estrogens but do, however, display some own specific mechanisms of signalization. Nevertheless, the ongoing efforts for the search of neuroprotective selective estrogen receptor-modulating molecules that do not trigger the adverse side effects of estrogens still deserve continuous investigation. More specifically, the putative neuronal survival-promoting and promnesic features of isoflavones, need further exploration to get full acceptance as beneficial natural substances against age-triggered alterations of general health including hippocampal-dependent memory decline.

In: Biochemistry Researcher Biographical Sketches ...
Editors: J. T. Walker and C. M. Jones
ISBN: 978-1-62081-308-9

Chapter 156

Prevention of Cardiovascular Diseases by Soy Protein and Isoflavones

***M. A. Rostagno*[*1], *N. Manchón*[1], *M. D`Arrigo*[1], *A. García- Lafuente*[1], *A. Villares*[1], *E. Guillamón,*[1,] *A. Ramos*[1] *and J. A. Martínez*[1,2]**

[1]Centro para la Calidad de los Alimentos,
Instituto Nacional de Investigación y Tecnología Agraria y Alimentaria (INIA),
Campus Universitario "Duques de Soria", Soria, Spain
[2]Universidad de Navarra, Dpto. Fisiología y Nutrición,
Edificio de Investigación, Pamplona, Spain

Research Summary

The use of functional foods in the diet is one of the latest trends in a strategy of intervention to reduce risk factors for the development of cardiovascular diseases. Soybeans and derived foods has attracted much attention of health professionals since its consumption is associated with a reduction of cholesterol levels in blood and other risk factors. In spite of the mechanisms and components responsible for the potential effect of soy foods for the prevention of cardiovascular diseases are not yet completely identified, the most recent studies suggest that several factors are inter-related, such as the functional composition of soybeans, the synergism between components in food, individual metabolism and genetics as well as other risk factors.

[*] Corresponding author: E-mail: rostagno.mauricio@inia.es, Telephone: (+34) 975 233204, Fax: (+34) 975 233205.

In: Biochemistry Researcher Biographical Sketches ... ISBN: 978-1-62081-308-9
Editors: J. T. Walker and C. M. Jones

Chapter 157

Effects of Processing on Isoflavone Content and Profile in Foodstuffs: A Review

Antonietta Baiano*
Department of Food Science, University of Foggia, Foggia, Italy,
Istituto per la Ricerca e le Applicazioni Biotecnologiche per la Sicurezza e la Valorizzazione dei Prodotti Tipici e di Qualità, University of Foggia, Foggia, Italy

Research Summary

Isoflavones, together with lignans, and coumestans represent the three classes of phytoestrogens, compounds that have an estrogen-like and antioxidant activity and are structurally similar to estradiol. Isoflavones are widespread in a variety of leguminous plants such as red clover, kudzu root and soy and are subjected to increasing attention due the possibility of use as potential therapeutic agents. In fact, several studies suggest that consumption of isoflavone-containing foods is associated with the lowest level of cholesterol in blood, and prevention of cardiovascular disease, prostate, colon and breast cancers, osteoporosis, and menopausal symptoms in people whose soy-rich diet allow to increase of 100 times the concentration of endogenous estrogens. Isoflavones are associated with the protein fraction of soybeans and are present under several chemical forms: the β-glucosides genistin, glycitin, daidzin, their 6”-*O*-malonyl-glucosides and a 6”-*O*-acetyl-glucosides and the aglycone forms genistein, daidzein and glycitein. Processing greatly affect the content of total isoflavones and the concentrations of each of their chemical forms. Among soy foods, the highest concentration of isoflavones is detectable in the whole soybeans whereas soy milk, tofu, tempeh, miso, fermented bean curd, bean paste and the so called “second-generation soyfoods” (soy hot dog, tempeh burger, soy noodles) show isoflavone contents ranging from 2 to 46% with respect to the whole soybeans. These reductions are mainly due to dilution of soy protein with other ingredients and heat treatments that also result in changes

* Corresponding author: Tel.: +39 (0)881 589117; fax: +39 (0)881 589308., e-mail address: a.baiano@unifg.it.

of the isoflavone profiles. In fact, the predominant isoflavones in soybeans and non-fermented soy products (soy protein, soymilk derivates, etc.) are malonyl glucosides, whereas fermented soy foods (miso, tempeh) contain mainly aglycones. These changes affect the metabolic activity of isoflavone-containing foods since the glycoside forms are biologically inactive and need to be cleaved by the β-glucosidases from intestine in the glucosyl residue and the biologically active aglycones.

This review gives an insight of the effects of processing and processing parameters (extraction, fermentation, pasta-making, addition of coagulants, addition of protease, temperature, pH) on isoflavone content, profile and antioxidant activity of soy foods.

In: Biochemistry Researcher Biographical Sketches ... ISBN: 978-1-62081-308-9
Editors: J. T. Walker and C. M. Jones

Chapter 158

Dietary Isoflavones and Intestinal Microbiota: Metabolism and Transformation into Bioactive Compounds

Maddalena Rossi[*], *Alberto Amaretti, Lucia Roncaglia, Alan Leonardi and Stefano Raimondi*
Dept. Chemistry, University of Modena and Reggio Emilia, Italy

Research Summary

Edible plants provide the human with hundreds of non-nutritional phytochemicals which are recognized as beneficial, such as isoflavones. Likewise other polyphenols, isoflavones may undergo extensive transformations during passage through human digestive tract, especially in the colon, where members of the complex commensal microbiota are capable to carry out synergistically a broad range of metabolic transformations affecting the fate and the biological activity of phytochemicals. Diverse bacterial species occurring in the large intestine hydrolyze the glucose conjugated forms of isoflavones, releasing the corresponding aglycones, which may undergo further microbial conversions (especially reductions) giving rise to a wide spectrum of isoflavone-derived compounds. The end-products of microbial transformations are subjected to substantial person-to-person variation, reflecting the impact of the colonic microbiota, since intestinal bacteria may greatly increase or compromise the biological activity of dietary isoflavones. In fact, certain reduction products (e.g. S-equol) possess superior antioxidant and estrogen-like activities, while others are precursors for C-ring cleavage and lead to isoflavone degradation. Extensive research has been performed to characterize the intestinal bacteria which are responsible for isoflavones transformations. In particular, many efforts are being carried out to identify single bacteria that convert the major

[*] Corresponding author: Dept. of Chemistry, University of Modena and Reggio Emilia, via Campi 183, 41125, Modena, Italy, tel. +39-059-2055567, fax. +39-059-373543, E-mail: maddalena.rossi@unimore.it.

soy isoflavone daidzein into S-equol, in the perspective to enrich soy products with such a valuable isoflavone-derived metabolite. Besides, diverse bacteria which are capable of deglycosylation, demethylation, and various reduction reactions are increasingly being identified. Nonetheless, only bifidobacteria and lactobacilli, which produce β-glucosidase that hydrolyze isoflavone glycosides into the corresponding aglycones, are currently exploited to improve the biological activity of soymilk.

In: Biochemistry Researcher Biographical Sketches ...
Editors: J. T. Walker and C. M. Jones
ISBN: 978-1-62081-308-9

Chapter 159

Bioactive Isoflavones: Implications in Cancer Treatment and Prevention

Alok Bhushan*, Vikas Bhardwaj, Cassandra Frandsen and James C. K. Lai
Department of Biomedical and Pharmaceutical Sciences, College of Pharmacy and ISU Biomedical Research Institute, Idaho State University, Pocatello, ID, US

Research Summary

Flavonoids are bioactive agents present in several plants that are used for food. These foods include legumes, soy, fruits, vegetables and different parts of plants. Vast amount of literature exits implicating the role of flavonoids in influencing changes in biological processes in cancer cells. The ability of flavonoids to induce changes in receptor activation, anti-estrogenic properties, tyrosine phosphorylation and cell signaling makes these agents important in cancer prevention and treatment and development of anticancer drugs. The antioxidant properties of flavonoids have also been extensively investigated. The above-mentioned cellular mechanisms are known to play critical roles in carcinogenesis, including metastasis and angiogenesis. Several isoflavones are in clinical trials and there has been epidemiological evidence for their role in cancer prevention and in decreasing incidence of several cancer types. Development of derivatives of isoflavones and associating these natural compounds with nanomaterials to give better efficacy and pharmacokinetic properties is some of the future areas of research which will have impact on incidence and treatment of an array of cancer types. This review will summarize some of the critical findings in the field and discuss the future directions in use of isoflavones in treatment and prevention of several selected cancer types.

Over half a million people in US die of cancer every year. Lung cancer is the major contributor of deaths due to smoking. The progression of cancer is a multi-step process and

* Corresponding author: Department of Bimedical and Pharmaceutical Sciences, College of Pharmacy, 970 S 5th Avenue, Idaho State University, Pocatello, ID 83209-8334, abhushan@pharmacy.isu.edu, Phone: 208-282-4408.

can be interrupted at several steps. The impact of consumption of isoflavones has been a topic of research and discussion for decades. To understand how isoflavones may block progression of cancer, it is necessary to discuss the process of carcinogenesis.

In: Biochemistry Researcher Biographical Sketches ...
Editors: J. T. Walker and C. M. Jones
ISBN: 978-1-62081-308-9

Chapter 160

Comparative in Vitro Effects of Soybean Isoflavones and Resveratrol on Human Osteoblastic Cells

Arancha R. Gortázar, Verónica Alonso and Pedro Esbrit *
Bone and Mineral Metabolism Laboratory,
Fundación Jiménez Díaz (Capio Group), Madrid, Spain

Research Summary

Estrogen, the main factor responsible for the maintenance of bone mass in women, inhibits bone resorption and increases bone density at least in part by interacting with osteoblasts. Estrogen replacement therapy effectively prevents bone loss after menopause, but also has adverse effects. Natural plant phytoestrogens, which have structural similarity to 17β-estradiol, are of growing interest by their potential role in osteoporosis prevention and treatment. In the present study, we compared several osteoblastic effects of the soy isoflavones genistein and daidzein, two well known phytoestrogens, and resveratrol, a phenolyc phytoestrogen abundant in grapes and wine. We used human osteosarcoma MG-63 cells, a characterized model for their nontransformed counterparts in a variety of studies on osteoblast function. We found that all these phytoestrogens, similarly to 17β-estradiol, at 100 nM, maximally increased alkaline phosphatase activity, as well as osteocalcin mRNA and osteoprotegerin (both mRNA and protein) expression. The protein levels of receptor activator of NF-κB ligand (RANKL) in MG-63 cell membrane extracts were also increased by either genistein or 17ß-estradiol but decreased by daidzein and resveratrol. However, the latter phytoestrogens, similarly to genistein and 17β-estradiol, decreased NF-κB-dependent interleukin-6 mRNA expression in MG-63 cells.

* Corresponding author: Laboratorio de Metabolismo Mineral y Óseo, Fundación Jiménez Díaz (Capio Group), Avda. Reyes Católicos, 2; 28040 Madrid, Spain, Phone: 34 91 5504894; Fax: 34 91 5498075, e-mail: pesbrit@fjd.es.

Our findings support previous findings in other nonhuman osteoblastic cell preparations, and indicate that relatively low concentrations of the main soy isoflavones and resveratrol display anabolic features apparently through estrogen receptors in human osteoblastic cells.

In: Biochemistry Researcher Biographical Sketches ... ISBN: 978-1-62081-308-9
Editors: J. T. Walker and C. M. Jones

Chapter 161

Health Effects of *Dioscorea Bulbifera* of Aerial-Potato (Yam)

*Noboru Motohashi**
Meiji Pharmaceutical University, Tokyo, Japan

Research Summary

Dioscorea spp. (yam) is a rich source of carbohydrate crop in the diet and also resource of different phytohormones. Among *Dioscorea* spp, especially, *Dioscorea bulbifera* of aerial-potato has widely been used as their traditional folklore medicines for their treatments such as thyroid disease, tumor, diretic disturbance and inflamation in the tropical and subtropical areas. The diverse compounds in *Dioscorea bulbifera* have been found and their functionality has been examined in various researches such as their antibacterial and antitumor activity. The purpose of this review paper is to focus and describe their compositions found and functional activities.

* Corresponding author: < noborumotohashi@jcom.home.ne.jp>

In: Biochemistry Researcher Biographical Sketches ... ISBN: 978-1-62081-308-9
Editors: J. T. Walker and C. M. Jones

Chapter 162

Effects of the Processing Techniques on Isoflavone Profiles

***Claudio L. Aguiar*[1], *Solange G. Canniatti-Brazaca*[1], *Yong K. Park*[2] *and Severino M. Alencar*[1]**
[1]Universidade de São Paulo, Escola Superior de Agricultura "Luiz de Queiroz", Piracicaba/SP, Brazil
[2]Faculdade de Engenharia de Alimentos, Universidade Estadual de Campinas (UNICAMP), Campinas, SP, Brazil

Research Summary

There is evidence that soy consumption benefits health by contributing to the prevention of chronic diseases due several factors, including isoflavones present in this legume. The isoflavones from soybean are known to have several biological properties such as estrogenic, antimicrobial and antitumoral (breast and prostate) activities, besides inhibiting the activity of enzymes linked to the cellular division. These properties are shown to be more accentuated in the aglycone than glycoside forms. Their presence in soy products have been reported with a number of beneficial health effects. In addition, both USA and Brazil are great soybean producer in the World. The analysis of isoflavone form contents into the soy products take an important role in the understanding of consumer of these products. In this way, with the germination process, the soybean has a positive effect in improving nutritional status including isoflavones. However, the isoflavone content varies with the period of germination. The process of germination showed no significant difference in amount of glycitin, genistein, malonyl daidzin, malonyl genistin, daidzein, glycitein and genistein, although theirs numerical reductions after germination. There were only differences in the amount of daidzein and malonyl glycitin between the samples. Some authors have been reported that the total concentration of isoflavones increased after 1 day of germination and then slowly decrease these concentrations. This increase and decrease observed in the content of isoflavones may be associated with the conversion of isoflavones to other flavonoids. The germination time of 48 hours not improved the amount of isoflavones. Different cultivars present several amount of isoflavones and forms of isoflavones. When the grains were cooked to be soft the malonyl form decreased, and aglycone and glycoside forms increased. It was observed that the heating treatment transformed the malonylglucosides into glucoside isoflavones. After heat treatment at 121°C for 30 min, nearly all malonyl isoflavones were converted into glucoside, but acetylisoflavones were not detected via RP-HPLC analysis. On the other hand, gamma radiation also promoted some reduction on isoflavone profiles, but this effect depends of humidity into soy products. Irradiation from ^{60}Co interfere at amount of isoflavones forms at doses up 2 kGy, glycosides and malonyl forms have tendency to down, but aglycones increase with radiation. For example, when soybean flour was studied, the gamma-radiation had a little effect (loss of 9.9%, w/w) in total isoflavone. However, soy isoflavones are likely to have the beneficial effects when found in aglycone forms in many processes will be proposed to obtain those aglycone isoflavones.

In: Biochemistry Researcher Biographical Sketches … ISBN: 978-1-62081-308-9
Editors: J. T. Walker and C. M. Jones

Chapter 163

Structural Studies and Mechanisms of Isoflavonoid Biosynthesis

Xiaoqiang Wang
Plant Biology Division, Samuel Roberts Noble Foundation,
Ardmore, OK, US

Research Summary

Isoflavonoids play important roles in plant defense for legumes, exhibit various health-promoting activities, and have attracted great interest from the pharmaceutical and nutraceutical industries as drug targets and dietary supplements. The isoflavonoids are identified primarily in leguminous plants. They are mainly synthesized through the central phenylpropanoid pathway and the specific isoflavonoid branch pathways. Many key enzymes in the central pathway have been elucidated structurally, providing essential insights into early stages of isoflavonoid biosynthesis. Significant progress has been made on structural studies of enzymes in the isoflavonoid branch pathway. Structures of two isoflavonoid-specific NADPH-dependent reductases revealed how enzymes specifically recognize isoflavonoids and catalyze stereo-specific reductions. Several isoflavonoid methyltransferases and glycosyltransferases were also characterized structurally. Bioactivities and functions of isoflavonoids are largely dictated by their methylation and glycosylation patterns. The structural analysis of these enzymes provides a basis for understanding the complex biosynthetic process and the selective substrate- and regio-specificities. Structure-based enzyme engineering is becoming a new strategy for synthesis of bioactive isoflavonoids.

In: Biochemistry Researcher Biographical Sketches ... ISBN: 978-1-62081-308-9
Editors: J. T. Walker and C. M. Jones

Chapter 164

Photophysics of Genistein and Biochanin a Isoflavones: Solvent Cage and Concentration Effects Studied by UV Visible Spectroscopy

K. Benthami[1], S. Ait Lyazidi[1,*], M. Haddad[1], B. Bennetau,[2] M. Choukrad and S. Shinkaruk[3]

[1]Laboratoire de Spectrométrie, des Matériaux et Archéomatériaux (LASMAR), URAC 11, Université Moulay Ismail, Faculté des Sciences, Meknès, Morocco
[2]Institut des Sciences Moléculaires (ISM), Université Bordeaux 1, Talence, France
[3]Ecole Nationale d'Ingénieurs des Travaux Agricoles (ENITA) de Bordeaux 1, Gradignan, France

Research Summary

Genistein and Biochanin (A) isoflavones are shown to exist in two different conformations which are the 90-degrees completely twisted geometry and the 50 degrees less twisted one, with respect to the C_3—$C_{1'}$ single bond. Specific interactions with the solvent cage as well as self association processes seem shifting the isoflavone from the perpendicular conformation towards the less twisted one. From the UV-visible photophysics investigations it's revealed, in both isoflavones, that monomeric species can't exist alone even at very low solute concentration (~10^{-6}M), the self association process occurs already in this concentration range. The theoretical simulation, using analytical atom-atom pair potential, predicts a self dimer in a slipped non sandwich, face to river, perpendicular structure.

[*] Corresponding author: E-mail : aitlyazidisaadia@yahoo.fr, phone: (+212) 666502265, Fax: (+212) 35536808.

In: Biochemistry Researcher Biographical Sketches ... ISBN: 978-1-62081-308-9
Editors: J. T. Walker and C. M. Jones

Chapter 165

Soy Isoflavone Genistein in Cancer Chemoprevention

M. F. Ullah, H. Zubair, H. Y. Khan and S. M. Hadi*

Department of Biochemistry, Faculty of Life Sciences, AMU, India

Research Summary

Soy isoflavones have been identified as dietary components having an important role in reducing the incidence of breast and prostate cancers in Asian countries. The isoflavone polyphenol genistein in soybean is considered to be a potent chemopreventive agent against cancer. Several studies, both *in vitro* and *in vivo*, have clearly shown that genistein is a promising therapeutic candidate for cancer chemoprevention as it is able to cause inhibition of human cancer growth via pathways that lead to apoptotic cell death. Genistein has pleiotropic molecular mechanisms of action affecting estrogen mediated signaling pathways, cell cycle regulation, apoptosis and cell growth signal cascades. Of particular interest is the observation that genistein has been found to induce apoptosis in cancer cell lines but not in normal cells. We have proposed a novel mechanism for the cytotoxic action of plant polyphenols including genistein, which has implications for its chemopreventive properties against cancer. We believe that such a mechanism better explains the preferential cytotoxicity of genistein towards cancer cells. In this review we discuss various mechanistic approaches that have been proposed to account for the chemopreventive properties of genistein.

* Corresponding author: E-mail: smhadi@vsnl.com, Tel: +91-571-2700741, Mobile: +919837266761.

In: Biochemistry Researcher Biographical Sketches ... ISBN: 978-1-62081-308-9
Editors: J. T. Walker and C. M. Jones

Chapter 166

Isoflavones in *Leguminosae*

***S. C. Cunha*[*,1], *T. Visnevschi-Necrasov*[#,2], *M. A. Faria*[†,1], *E. Nunes*[‡,2] and *M. B. P. P. Oliveira*[¶,1]**

[1]REQUIMTE - Serviço de Bromatologia, Faculty of Pharmacy, University of Porto, Rua Aníbal Cunha, Porto, Portugal
[2]CIBIO-ICETA, Faculty of Science of University of Porto, Vairão, Portugal

Research Summary

Isoflavones are a class of plant secondary metabolites, related to flavonoids which are most commonly known for their antioxidant activity. Flavonoids are a type of polyphenol derived from the shikimate pathway, responsible for plant defense and flower color. Plants use isoflavones and their derivatives as phytoalexin compounds to ward off disease-causing pathogenic fungi and other microbes. Isoflavones are produced via a branch of the general phenylpropanoid pathway that produces flavonoid compounds in higher plants.

Isoflavones are produced almost exclusively by the members of the Fabaceae/Leguminosae (bean) family. Among the members of the Fabaceae family that contain significant quantities of isoflavones, Glycine max (soybean) and soy products assume particular importance as source of isoflavones in human nutrition. However, the allergenic nature of many soy products for the people and the difficulty in obtaining genetically modified organism (GMO)-free soybeans are causing experts in industry to look for

* REQUIMTE - Serviço de Bromatologia, Faculty of Pharmacy, University of Porto, Rua Aníbal Cunha, 164, 4099-030 Porto, Tel: +351222078910, Fax: +351222003977, Email: sara.cunha@ff.up.pt, Both authors contribute equally for this work

CIBIO-ICETA, Faculty of Science of University of Porto, R. Padre Armando Quintas 4485-661 Vairão, Portugal, Tel: +351252660402, Fax: +351252661780, Email: tativis74@gmail.com, Both authors contribute equally for this work

† REQUIMTE - Serviço de Bromatologia, Faculty of Pharmacy, University of Porto, Rua Aníbal Cunha, 164, 4099-030 Porto , Tel: +351222078910, Fax: +351222003977, Email: mfaria@ff.up.pt

‡ CIBIO-ICETA, Faculty of Science of University of Porto, R. Padre Armando Quintas 4485-661 Vairão, Portugal, Tel: +351252660402, Fax: +351252661780, Email: enunes@mail.icav.up.pt

¶ REQUIMTE - Serviço de Bromatologia, Faculty of Pharmacy, University of Porto, Rua Aníbal Cunha, 164, 4099-030 Porto,Tel: +351222078927, Fax: +351222003977, Email: beatoliv@ff.up.pt

alternative sources of isoflavones. Thus, other legumes including Trifolium pratense (red clover) have recently received considerable attention due to its high isoflavones content in which has been reported to be 2-10 times higher than in soybean seeds, and because its wide adaptaptation to different agro-ecological regions of the world. Red clover contains genistein and daidzein as soybean and, more interestingly, it contains significantly higher concentrations of biochanin A and formononetin, methylderivatives of genistein and daidzein.

In: Biochemistry Researcher Biographical Sketches ... ISBN: 978-1-62081-308-9
Editors: J. T. Walker and C. M. Jones

Chapter 167

Diffuse Neutron Scattering Study of Correlation Effects among Thermal Displacements of Atoms

Takashi Sakuma[1*], Xianglian[2], Hiroyuki Uehara[1] and Saumya R. Mohapatra[1]
[1]Institute of Applied Beam Science, Ibaraki University, Mito, Japan
[2]College of Physics and Electronics Information, Inner Mongolia University for the Nationalities, Tongliao, China

Research Summary

Diffuse scatterings data contain information about static and dynamic disorders in crystals. Neutron diffuse scattering measurements from ionic crystals, covalent crystals and metal crystals are investigated at low and room temperatures by angular-dispersive and energy-dispersive scattering methods. The intensity of the diffuse scattering changes with temperature. The oscillatory forms of the diffuse scattering are observed even from ordered crystals at room temperature. The observed diffuse scattering intensities are analyzed by including the correlation effects among thermal displacements of atoms. The values of correlation effects decrease with decrease of temperature. For most of the materials, the value of correlation effects among first nearest neighboring atoms is observed to be ~ 0.7 at room temperature. The values of correlation effects decrease rapidly with the increase of inter-atomic distances. The correlation effects do not depend much on the type of the crystal structure and crystal binding. From the parameters of Debye-Waller temperature factor and the values of correlation effects, the inter-atomic force constants among first, second and third nearest neighboring atoms are determined.

* Corresponding author: e-mail address: sakuma@mx.ibaraki.ac.jp. Tel.: +81-29-228-8357; Fax: +81-29-228-8357.

A rough estimation of phonon dispersion relation has been tried from the inter-atomic force constants and crystal structure using computer simulation. The diffuse background scattering measurement is found to be very useful in optimizing annealing process during preparation of lithium-based metal oxides.

In: Biochemistry Researcher Biographical Sketches ... ISBN: 978-1-62081-308-9
Editors: J. T. Walker and C. M. Jones

Chapter 168

Small Angle Scattering Analysis of Virus-like Particles for Biomedical Diagnostic Assays

Susan Krueger[1], Janet L. Huie[2] and Deborah A. Kuzmanovic*[3]
[1]NIST Center for Neutron Research, NIST, Gaithersburg, MD, US
[2]Sciencenter, Ithaca, NY, US
[3]University of Michigan, Biophysics and the Department of Molecular, Cellular, and Developmental Biology, Ann Arbor, MI, US

Research Summary

Technical advances in biotechnology have made the production of genetically engineered biomaterials commonplace. Many artificially produced biomolecules retain native binding to protein or nucleic acid targets, assemble into higher order complexes, and perform native catalytic functions.

Virus-based tools for biomedical purposes are of particular interest. Virus coat proteins can spontaneously assemble into intact virus-like particles (VLPs), with many structural features of the native virus. Specifically, the VLP displays surface antigens for host recognition, and its protective shell prevents damage to and acts as a delivery agent for its contents. Unlike native viruses, VLPs are noninfectious and usually lack nucleic acid content. These features have been used to advantage to produce vaccines such as for human papillomavirus (HPV). However, VLP vaccine development would be greatly served by a facile, high resolution method for physical characterization of VLPs in the form to be administered (i.e., in solution).

VLPs are also ideally suited as standards for corresponding infectious viral particles in clinical diagnostic assays. In particular, VLP carriers of short nucleic acid sequences show

*Corresponding Author. University of Michigan, Biophysics and the Department of Molecular, Cellular, and Developmental Biology, 930 N. University, 3039 Chemistry Bldg, Ann Arbor, MI 48109, Tel: (734)-647-9121, Fax: (734)-764-3323, Email: kuzman@umich.edu.

great promise as quantitative reference standards in DNA and RNA diagnostic assays. However, as for VLP vaccine development, the use of nucleic acid-filled VLPs in diagnostic testing has been hampered by a lack of physical techniques to easily measure VLP physical properties in solution.

Small-angle neutron scattering (SANS) and small-angle x-ray scattering (SAXS) are novel approaches to physical characterization of VLPs for vaccine and diagnostic standard development. SANS offers the unique capability of studying both the coat protein and the RNA or DNA of interest as separate entities, within a VLP containing both components. Both SAXS and SANS measurements are performed using VLPs in solution. This chapter examines the unique power and utility of scattering techniques for structural characterization of native viral particles, nucleic acid-free VLPs (reconstituted capsids), and recombinant VLPs carrying a foreign RNA of interest. In addition, the analysis of SANS and SAXS results will be demonstrated for modeling the structure of the coat protein and encapsulated RNA (required for vaccine development) in solution, as well as for quantifying the amount of RNA in a VLP. Quantification of nucleic acid content is critical for assessment of whether recombinant VLPs meet the requirements for a standard biomedical assay.

In: Biochemistry Researcher Biographical Sketches ...
Editors: J. T. Walker and C. M. Jones
ISBN: 978-1-62081-308-9

Chapter 169

Small Angles Neutrons Scattering Application in Biological Materials

Adel Aschi and Abdelhafidh Gharbi
Lab. De Physique de la Matiere Molle, Department de Physique Faculte des Sciences de Tunis, Campus Universitaire, Tunisia

Research Summary

In this part of book, we presented the different steps to study the structure and properties of different proteins by small angles neutrons scattering (SANS). We have shown in this case, how to treat the neutron scattering spectra i.e the corrections of spectra. The form and structure factors used to determine respectively the structure and interactions effects. The figures presented are good witnesses and examples for these.

In: Biochemistry Researcher Biographical Sketches … ISBN: 978-1-62081-308-9
Editors: J. T. Walker and C. M. Jones

Chapter 170

Neutron Scattering in Skin Research

Jarmila Zbytovská[1*] and Kateřina Vávrová[2]
[1]Department of Pharmaceutical Technology, Charles University in Prague, Faculty of Pharmacy, Hradec Králové, Czech Republic
[2]Centre for New Antivirals and Antineoplastics, Department of Inorganic and Organic Chemistry, Charles University in Prague, Faculty of Pharmacy, Hradec Králové, Czech Republic

Research Summary

The mammalian skin plays a major role in the protection of the body against pathogen threats from the environment. The unique skin barrier properties are ensured by the outermost skin layer, the *stratum corneum* (SC). The SC consists of corneocytes which are embedded in a lipid matrix consisting of ceramides, free fatty acids, cholesterol, and its derivatives. The SC lipids are organized in lipid membranes arranged into a lamellar structure. Due to the special physicochemical properties of ceramides, these membranes are extremely rigid and, therefore, very poorly permeable. A number of skin diseases, namely atopic dermatitis, psoriasis or recessive X-linked ichthyosis, are related to an impaired structure of the SC lipid membrane. On the other hand, a reversible decrease in the skin barrier function connected with the changes in the SC lipid membrane organisation is often required in the transdermal drug delivery. These facts encourage looking for new approaches towards investigating the internal membrane arrangement of the SC lipid matrix.

Recently, several neutron scattering techniques have been applied in this context. Above all, the neutron scattering on unilamellar vesicles and the neutron diffraction on oriented multilamellar lipid films allow describing the organisation of SC lipid membrane models on the molecular level. These promising results open new opportunities in the skin research.

The present chapter aims to review the recent experience with the neutron scattering methods concerning the investigation of the SC lipid membrane structure.

* Corresponding author: Faculty of Pharmacy, Charles University in Prague, Department of Pharmaceutical Technology, Heyrovského 1203, 50005 Hradec Králové, Czech Republic, Tel. No.: +420 495067383, E-mail address: Jarmila.Zbytovska@faf.cuni.cz.

In: Biochemistry Researcher Biographical Sketches ...
Editors: J. T. Walker and C. M. Jones
ISBN: 978-1-62081-308-9

Chapter 171

Precipitate Analysis of Microalloyed Steels using Small Angle Neutron Scattering

J. Barry Wiskel
Dept. of Chemical and Materials Engineering, University of Alberta,
Edmonton, Alberta, Canada

Research Summary

Precipitates (2nd phase particles) are a major strengthening mechanism in many types of steels. In particular, it is the size (and distribution) and volume fraction of very fine precipitates (≈<10 nm) that contribute significantly to the strengthening of these materials. This important precipitate size range coincides well with the precipitate analysis capabilities of Small Angle Neutron Scattering (SANS). The major challenges in characterizing precipitates in microalloyed steels are the relatively low volume fraction of nano size precipitates (<0.20wt%) and both the wide range of composition and sizes of precipitates that can form during the TMCP process. Given these challenges, characterizing precipitates in these types of steels requires a diversity of techniques including TEM, SEM, EDX, XRD and SANS. This paper will detail the analysis of nano sized precipitates in microalloyed steels using SANS. The precipitate data (e.g. size and distribution of the nano sized precipitates) measured with SANS was compared with other measurements techniques (TEM and XRD) for validation. In addition, the importance of the SANS precipitate data generated and how it can be used to correlate nano precipitates size and volume fraction with steel processing parameters will also be presented. This relationship can be used to identify steel processing conditions that would enhance fine precipitate evolution and ultimately the strength of the steel.

In: Biochemistry Researcher Biographical Sketches … ISBN: 978-1-62081-308-9
Editors: J. T. Walker and C. M. Jones

Chapter 172

Small Angle Neutron Scattering from Proteinsin Solution

Maria Grazia Ortore*

Dipartimento SAIFET and CNISM, Universit`a Politecnica delle Marche, Ancona, Italy

Research Summary

The study of biomolecules in solution is quite important in order to reproduce environmental conditions quite resembling those existing in-vivo. Small Angle Neutron Scattering (SANS) from proteins in solution provides information concerning size and shape, over a large length scale. Performance of SANS experiments on samples at different deuteration grades enables a deeper knowledge both of protein structure and of protein solvation shell. The chance to perform numerical simulation before car- rying on a SANS experiment can lead the user to a wiser choice of the experiment preparation and greatly improve the structural resolution. A brief discussion about the scattering theory, isotopic substitution and thermodynamic equilibria between water and cosolvent molecules in the protein solvation shell, will be presented. This chap- ter will discuss theory and practical aspects with protein conditions in order to guide potential users to successfully apply SANS experiments to answer peculiar biological questions involving proteins in solution.

* Tel.: (+39)071-2204664,Fax:(+39)071-2204605.

In: Biochemistry Researcher Biographical Sketches ... ISBN: 978-1-62081-308-9
Editors: J. T. Walker and C. M. Jones

Chapter 173

Neutron and X-Ray Diffraction Studies of Nanoparticles Confined within Porous Media

I. V. Golosovsky [*]
Petersburg Nuclear Physics Institute, St. Petersburg, Russia

The practical needs and technology achievements make available large diversity of nanos- tructured systems. They can be arranged into several more or less definite classes. First, there are isolated nanoparticles, usually weakly interacting, sometimes immersed into any suspension (liquid), or/not covered with some coating. They are of interest from fundamen- tal point of view as initial blocs of any nanostructured magnetic systems. From practical point of view they are of importance for medical applications. Another class is multicom- ponent systems. One can divide them onto two sub-classes: composite systems, where nanoparticles are embedded (synthesized) within different type of porous media, for ex- ample, within a porous glass. Another class is heterogeneous systems, layered [1, 2] or "core-shell" structures [3, 4] with properties which are engineered by controlling the mi- crostructure and chemical profile of the layers. All nanostructured materials can be treated as ordinary compounds, which are synthe- sized within artificial boundaries, in the so-called conditions of "restricted geometry". All these objects have common peculiarities. First, their dimensions are comparable to the cor- relation lengths of atomic, magnetic and other characteristic interactions. Such restrictions lead to a change in the nature of the phase transition, sometimes to its complete suppression. Secondly, the quantity of atoms on the surface, which exist in the conditions of violation of local symmetry, is comparable with the total number of atoms in a system that significantly modifies physical properties of nanostructured system. Therefore, in such systems the usual consideration when surface atoms are simply ignored, is not valid. The subject of the presented chapter is diffraction studies of nanocomposites, mate- rials which obtained by the chemical synthesis (or other methods) of oxides, metals and other compounds inside porous media. The study of the

[*] E-mail address:golosov@mail.pnpi.spb.ru.

physical properties of confined nanoparticles has become a very active field of research during the last decade. The reason to investigate such materials is fundamental since the confined geometry and the influence of the surface yields unusual properties as compared with the bulk. Confined nanoparticles also result in new applications, for example, in the field of catalysis, high-density magnetic memory, etc.

In: Biochemistry Researcher Biographical Sketches … ISBN: 978-1-62081-308-9
Editors: J. T. Walker and C. M. Jones

Chapter 174

Methods in Protein Structure and Stability Analysis: Conformational Stability, Size, Shape and Surface of Protein Molecules

Vladimir N. Uversky and Eugene A. Permyakov

Research Summary

Protein research is a frontier field in science. Proteins are widely distributed in plants and animals and are the principal constituents of the protoplasm of all cells, and consist essentially of combinations of a-amino acids in peptide linkages. Twenty different amino acids are commonly found in proteins, and serve as enzymes, structural elements, hormones, immunoglobulins, etc., and are involved throughout the body, and in photosynthesis. This new book gathers new leading-edge research from throughout the world in this exciting and exploding field of research.

In: Biochemistry Researcher Biographical Sketches ... ISBN: 978-1-62081-308-9
Editors: J. T. Walker and C. M. Jones

Chapter 175

Neutron Scattering from Solutions of Biological Macromolecules*

M. W. Roessle and D. I. Svergun
EMBL-Outstation Hamburg, Hamburg, Germany

Research Summary

Small-angle scattering (SAS) of X-rays and neutrons is a powerful method for the analysis of biological macromolecules in solution. The method allows one to study low resolution structure of native particles in nearly physiological environments and to analyze dynamic processes like complex formation, assembly and folding. Thanks to recent progress in SAS instrumentation and novel analysis methods, which substantially improved the resolution and reliability of the structural models, the method has attracted renewed interest of biologists and biochemists. In this chapter devoted to biological neutron scattering (SANS) we cover basics of the neutron scattering theory, SANS instrumentation and modern methods used in data analysis and modeling. Differences between X-ray and neutron scattering and advantages/shortcomings of the two techniques are demonstrated. The major advantage of SANS, *i.e.* possibility of contrast variation and specific labeling due to isotope hydrogen/deuterium exchange is comprehensively discussed. Recent applications of SANS to proteins and nucleoprotein complexes are reviewed and a test case, contrast variation analysis of a GroEL/GroES chaperonin system using modern data interpretation methods, is presented in detail.

* A version of this chapter was also published in *Methods in Protein Structure and Stability Analysis: Conformational Stability, Size, Shape and Surface of Protein Molecules*, edited by Vladimir N. Uversky and Eugene A. Permyakov, Nova Science Publishers. It was submitted for appropriate modifications in an effort to encourage wider dissemination of research.

In: Biochemistry Researcher Biographical Sketches ... ISBN: 978-1-62081-308-9
Editors: J. T. Walker and C. M. Jones

Chapter 176

Elastic, Quasi Elastic and Inelastic Neutron Scattering Studies on Hydrogen-Bonded Systems of Biophysical Interest*

***S. Magazù and F. Migliardo*†**
Dipartimento di Fisica, Università di Messina, Messina, Italy

Research Summary

Neutron scattering has been demonstrated to be a powerful tool for investigating physical and chemical mechanisms of biophysical processes. The present paper shows Elastic (ENS), Quasi Elastic (QENS) and Inelastic Neutron Scattering (INS) findings on a class of hydrogen-bonded systems of biophysical interest, such as homologues disaccharides (i. e. trehalose, maltose and sucrose)/water mixtures as a function of temperature and concentration.

The aim of the present work is to give an overview of dynamical properties of the investigated systems in order to characterise the dynamical transition, the diffusion and the vibrational modes of disaccharides in solution and to link the observed features to their bioprotective effectiveness. The role of hydrogen bond and fragility in these bioprotectant systems is discussed in detail. The obtained findings highlight a strong destructuring effect on the hydrogen-bonded network and a marked slowing down effect on the dynamics of water by disaccharides together with a considerable cryptobiotic action and can justify the superior capability of trehalose in respect to maltose and sucrose to encapsulate biostructures in a more rigid and protective environment.

---

* A version of this chapter was also published in Chemical Physics Research Journal, Volume 1, Issue 2-3, Nova Science Publishers. It was submitted for appropriate modifications in an effort to encourage wider dissemination of research.

† E-mail address: fmigliardo@unime.it. Tel.: +39 0906765019. Fax: +39 090395004 Federica Migliardo, Dipartimento di Fisica, Università di Messina, C.da Papardo, S.ta Sperone 31, P.O. Box 55, 98166 S. Agata, Messina, Italy. (Corresponding author)

In: Biochemistry Researcher Biographical Sketches … ISBN: 978-1-62081-308-9
Editors: J. T. Walker and C. M. Jones

Chapter 177

Milk Oligosaccharides

***Tadasu Urashima*[1]*, Motomitsu Kitaoka*[2]*, Takashi Terabayashi*[3]*, Kenji Fukuda*[1]*, Masao Ohnishi*[4] *and Akira Kobata*[5]**

[1]Graduate School of Animal Hygiene,
Obihiro University of Agriculture and Veterinary Medicine,
Obihiro, Hokkaido, Japan
[2]National Food Research Institute, National Agriculture
and Food Research Organization, Tsukuba, Ibaraki, Japan
[3]Department of Chemistry, Faculty of Science,
Kitasato University, Sagamihara, Kanagawa, Japan
[4]Department of Food Science, Obihiro University of Agriculture &
Veterinary Medicine, Obihiro, Hokkaido, Japan
[5]Noguchi Institute, Itabashi, Tokyo, Japan

Research Summary

Most human milk oligosaccharides (HMO) are resistant to digestion and absorption within the small intestine and reach the infant colon, where they can act as prebiotics that stimulate the growth of beneficial microorganisms such as various species of Bifidobacterium. They can also act as receptor analogues that inhibit the attachment of harmful microorganisms to the infant's colonic mucosa. A small part of the milk oligosaccharides are absorbed intact into the circulation and it has been hypothesized that these may act as immunomodulators. It is generally believed that bovine milk oligosaccharides are not absorbed by human adults or infants, thus making them available to be utilized as prebiotics or anti-infection materials. This book presents and discusses research on milk oligosaccharides.

In: Biochemistry Researcher Biographical Sketches ... ISBN: 978-1-62081-308-9
Editors: J. T. Walker and C. M. Jones

Chapter 178

Prebiotic Oligosaccharides: Origins and Production, Health Benefits and Commercial Applications

Santad Wichienchot [*] ***and Pavinee Chinachoti***
Nutraceutical and Functional Food Research and Development Center,
Faculty of Agro-Industry, Prince of Songkla University, Hat Yai, Songkhla, Thailand

Research Summary

In recent years, oligosaccharides and their derivatives have become useful for health applications in various fields because of their specific biological activities. In the food industries, several oligosaccharides have received increasing attention as key components for functional foods and nutraceutical products. Prebiotics are non-digestible oligosaccharides which have been shown to have properties that can modulate gastrointestinal problems and improve general health and well being. The benefits of prebiotic include relief of constipation, reduced risk of colon cancer, inhibition of pathogens in gastrointestinal tract, increased minerals absorption, immune modulation, short chain fatty acid and vitamin production, reduced blood cholesterol and lipids and improved microbial balance in the gut. Prebiotics are distinguished from other dietary fibers due to its ability to selectively promote fermentation by bifidobacteria and/or lactobacilli within the gut. The "prebiotics" was named by Gibson and Roberfroid in 1995 as "a non digestible food ingredient that beneficially affects the host by selectively stimulating the growth and/or activity of one or a limited number of bacteria in the colon, that can improve the host health". It was redefined several times until 2007 when FAO defined as a "non-viable food component that confers a health benefit on the host associated with modulation of the microflora.

Currently, commercial prebiotics are non-digestible but fermentable oligosaccharides which are consisted of sugar moieties between 2 and 10 saccharide units, with an exception of inulin and beta-glucans which are polysaccharides. Prebiotic oligosaccharides can be obtained

[*] E-mail: santad.w@psu.ac.th.

from natural sources such as chicory and artichoke and produced through chemical synthesis (lactulose) or by enzymatic synthesis (hydrolysis or transfer reaction). Currently commercialized prebiotic oligosaccharides for food use include fructo-oligosaccharides, galacto-oligosaccharides, lactosucrose, isomalto-oligosaccharides, gentio-oligosaccharides, xylo-oligosaccharides and soybean oligosaccharides (whereas lactulose has been used as a laxative). Recommendation on prebiotics consumption is 3-10 g per day. Effectiveness on each type of prebiotic varies; for example, lactulose, lactosucrose and xylo-oligosaccharides have been reported more effective at low doses. Prebiotics are popular food ingredients used in functional foods and nutraceutical products such as low calories and mind sweetness for weight control or diabetes and for well being of gut health. Combination of prebiotics and probiotics (synbiotic products) offer synergistic benefits to the GI tract, such as in infant powder formula, beverage, yoghurt and dairy products. Functional claims of prebiotic depend on local law of each country, however, most country accepted as a source of dietary fiber.

In: Biochemistry Researcher Biographical Sketches ...
Editors: J. T. Walker and C. M. Jones
ISBN: 978-1-62081-308-9

Chapter 179

Glycoside Hydrolases from Hyperthermophiles: Structure, Function and Exploitation in Oligosaccharide Synthesis

Beatrice Cobucci-Ponzano**, Mosè Rossi and Marco Moracci
Institute of Protein Biochemistry – CNR, Naples, Italy

Research Summary

Hyperthermophilic microorganisms thrives at temperatures higher than 80°C and proteins and enzymes extracted from these sources are optimally stable and active in the presence of temperatures close to the boiling point of water and of other denaturants, i.e. chaotropic agents, pH, organic solvents, detergents, etc. Therefore, hyperstable enzymes are considered attractive alternatives in biocatalysis and in chemo-enzymatic synthesis. In addition, the molecular bases of the extreme stability to heat and to the ability to work optimally at high temperatures are not completely understood and intrigued biochemists, enzymologists, and biophysics in the last twenty years. In particular, hyperstable glycosidases, enzymes catalysing the hydrolysis of *O*- and *N*-glycosidic bonds, have been studied in detail as they are simple model systems promoting single-substrate reactions, and, more importantly, can be exploited for the enzymatic synthesis of oligosaccharides. The importance of these molecules increased enormously in recent years for their potential application in biomedicine. Hyperstable glycosidases, working in transglycosylation mode, can be excellent alternatives to the classical chemical methods helping in the control of regio- and stereoselectivity as conventional enzymes, but also resisting to the organics used in chemical synthesis. We will review here recent advances in the isolation and characterization of glycosidases from hyperthermophilic microorganisms and the methods used for their application in oligosaccharide synthesis.

* *To whom the correspondence should be addressed: Tel. +39-081-6132564; Fax +39-081-6132277; E-mail b.cobucciponzano@ibp.cnr.it.

In: Biochemistry Researcher Biographical Sketches … ISBN: 978-1-62081-308-9
Editors: J. T. Walker and C. M. Jones

Chapter 180

Enzymatic Synthesis of Linear, Cyclic and Complex Type Oligosaccharides

***Piamsook Pongsawasdi*[1] *and Kazuo Ito*[2]**
[1]Department of Biochemistry, Faculty of Science,
Chulalongkorn University,
Bangkok, Thailand
[2]Laboratory of Enzyme Chemistry, Graduate School of Science,
Osaka City University, Osaka, Japan

Research Summary

For oligosaccharide synthesis, an enzymatic process is preferred over the complicated multi-step reactions of chemical synthesis. The advantages of the enzymatic process are the regio- and stereo-selectivity, potential for large-scale synthesis, and also the environmental-friendly synthesis conditions. Two main classes of enzymes, the glycosidases and the glycosyltransferases, are exploited in the common approach for oligosaccharide synthesis via reverse hydrolysis and transglycosylation reactions. The glycosynthase class has recently been developed from specific mutations of glycosidases through site-directed mutagenesis to minimize the natural hydrolysis activity and thereby boost up the synthetic activity of the enzyme. In this chapter, oligosaccharides synthesized by enzymes are classified into three main types, the linear, cyclic and complex oligosaccharides. Our work on three major enzymes, comprised of two glycosyltransferases and one glycosidase that are capable of synthesizing oligosaccharides is outlined. The use of bacterial cyclodextrin glycosyltransferase and amylomaltase in intermolecular transglucosylation to produce functional linear oligosaccharides is presented. In addition, these two enzymes, through the intramolecular transglucosylation reaction, are used for the synthesis of small and large cycloamyloses (cyclodextrins) that are widely used as stabilizers and solubilizers. For the complex-type oligosaccharides, the transglycosylation activity of endo-β-*N*-acetylglucosaminidase to yield the high mannose-GlcNAc-Glc oligosaccharides with potential applications in biological functions, such as therapeutic agents, is discussed. The chapter ends with concluding remarks on the commercially available oligosaccharides, with the emphasis on the significance of oligosaccharides and oligosaccharide-producing enzymes and their promising future.

In: Biochemistry Researcher Biographical Sketches ... ISBN: 978-1-62081-308-9
Editors: J. T. Walker and C. M. Jones

Chapter 181

Precursor N-Linked Oligosaccharides as Codes for Glycoprotein Folding Status

Ron Benyair and Gerardo Z. Lederkremer*
Department of Cell Research and Immunology, George Wise Faculty of Life Sciences, Tel Aviv University, Tel Aviv, Israel

Research Summary

The majority of eukaryotes share a common N-linked oligosaccharide precursor, $Glc_3Man_9GlcNAc_2$, which is transferred to proteins during glycoprotein biosynthesis. This precursor is then sequentially processed in the endoplasmic reticulum (ER), creating a series of oligosaccharide structures that are recognized as codes by specific lectins and that inform of the folding status of the glycoprotein. For example, the chaperones/ lectins calnexin and calreticulin bind to monoglucosylated oligosaccharides after the excision of the two terminal glucose residues by glucosidases I and II. The last glucose and sometimes one or even two mannose residues are excised, and the resulting structures $Man_{7\text{-}9}GlcNAc_2$ are recognized by the lectins ERGIC53, VIP36 and others that are involved in transport of glycoproteins to the Golgi. These lectins associate with properly folded glycoproteins that can exit the ER. In contrast, N-linked oligosaccharides on misfolded glycoproteins are more extensively trimmed to $Man_{5\text{-}6}GlcNAc_2$. A high local concentration of ER mannosidase I in an ER-derived quality control compartment is mainly responsible for this trimming, with the possible participation of other mannosidases. $Man_{5\text{-}6}GlcNAc_2$ oligosaccharides are then recognized by the lectins OS9 and XTP3-B that target the misfolded glycoprotein for ER-associated degradation.

* Email: gerardo@post.tau.ac.il
Phone: 972-3-640-9239. FAX: 972-3-642-2046.

In: Biochemistry Researcher Biographical Sketches … ISBN: 978-1-62081-308-9
Editors: J. T. Walker and C. M. Jones

Chapter 182

Oligosaccharides from Sucrose via Glycansucrases

Gregory L. Côté[*,1]
Renewable Product Technology Research Unit
National Center for Agricultural Utilization Research
Agricultural Research Service
United States Department of Agriculture, IL, US[#]

Research Summary

Glycansucrases are a class of microbial enzymes that polymerize either the fructosyl or the glucosyl moiety of sucrose to give β-D-fructans or α-D-glucans. They are also capable of transferring fructosyl or glucosyl units to acceptor molecules to yield oligosaccharides. Although the glycosyl donor specificity is limited to sucrose and related sugars, the acceptor specificity is very broad, and includes numerous carbohydrates as well as non-carbohydrate molecules. This chapter describes the enzymes and the variety of oligosaccharide structures that result from their acceptor reactions. Applications can range from modified drugs to food ingredients. Although very few of the products have thus far been commercialized, the potential is being actively studied and shows great promise.

[*] Email: greg.cote@ars.usda.gov. Telephone: 309-681-6319; Fax 309-681-6040
[#] Mention of a trade name, proprietary product, or specific equipment does not constitute a guarantee or warranty by the United States Department of Agriculture and does not imply its approval to the exclusion of other products that may be suitable.

In: Biochemistry Researcher Biographical Sketches ... ISBN: 978-1-62081-308-9
Editors: J. T. Walker and C. M. Jones

Chapter 183

Electrospray Mass Spectrometry of Oligosaccharides of Plant Origin

Maria do Rosário M. Domingues[1], *Fernando M. Nunes*[2] *and Manuel A. Coimbra*[1]
[1]Departamento de Química, Universidade de Aveiro, Aveiro, Portugal
[2]Departamento de Química, Universidade de Trás-os-Montes e Alto Douro, Vila Real, Portugal

Research Summary

Carbohydrates are widely distributed in nature, being responsible for different functions in almost all living organisms. They are present in the living systems as polysaccharides, glycoproteins, and glycolipids. Their properties and applications are dependent on their detailed structure, namely, their composition in monosaccharides, type of linkages, branching, and anomeric configurations. Mass spectrometry (MS) is a useful tool for the study of carbohydrates, especially when soft ionization methods are used. For the analysis of oligosaccharides, electrospray mass spectrometry (ESI-MS) and tandem mass spectrometry (ESI-MS^n) has been elected in the last recent years as the method of choice, having the advantage of using underivatised oligomers, even when present in mixtures and with low abundance.

The analysis of polysaccharides in their native form is difficult due to their high molecular weight. In order to overcome this, polysaccharides are usually cleaved into oligosaccharides.

In the present work the structural features of glucuronoxylans, galactomannans, and type-II arabinogalactans achieved by the analysis by ESI-MS of their oligosaccharides are discussed. In addition, it is shown how can the anomeric configurations of reducing disaccharides of glucose using quadrupole time-of-flight (Q-TOF2), linear ion trap (LIT), and triple quadrupole (QqQ) mass spectrometers be obtained.

In: Biochemistry Researcher Biographical Sketches … ISBN: 978-1-62081-308-9
Editors: J. T. Walker and C. M. Jones

Chapter 184

Biologically Active Oligosaccharide Functions in Plant Cell: Updates and Prospects

Olga A. Zabotina[1] and Aleksey I. Zabotin[2]
[1]Department of Biochemistry, Biophysics and Molecular Biology,
Iowa State University, Ames, IA, US;
[2]Institute of Biochemistry and Biophysics,
Russian Academy of Sciences, Kazan, Russia

Research Summary

Biologically active oligosaccharides, referred to as oligosaccharins, are the specific group of complex carbohydrates that function in plant cells as molecular signals. Indeed, oligosaccharins participate in the regulation of growth, development, and survival in different environmental conditions. Significant research has helped to shed light on the oligosaccharin concept and to make progress in this field. For instance, various oligosaccharins have been structurally characterized, and for some of them, specific receptors have been discovered. Separate signaling events have also been elucidated. Currently, a number of scientific articles and comprehensive reviews present a large body of evidence supporting the idea of oligosaccharin existence and function. This review provides a brief overview of most recent information obtained about oligosaccharin functions, signal perception, their possible origins, and movement within the plant. The discussion is focused only on oligosaccharins that function in plant cells, either originating from the plant cell wall or from the fungal wall. Please note that there are a number of other biologically active carbohydrate-containing molecules, such as fragments of glycoproteins or lipo-oligosaccharides, which are not discussed in this review. Despite gaps in our knowledge of oligosaccharin function, there is no doubt that this group of signaling molecules plays distinct roles in plant cells. The identification of new structures and activities of oligosaccharins in a broader range of organisms, together with the elucidation of their signal transduction pathways, will broaden our understanding of the important roles of these molecules in different aspects of plant life and will yield valuable insight into how cells progress through developmental programs and respond to environmental changes.

In: Biochemistry Researcher Biographical Sketches ...
Editors: J. T. Walker and C. M. Jones

ISBN: 978-1-62081-308-9

Chapter 185

The Mannan Oligosaccharides in Aquaculture

Huynh Minh Sang*[*] *and Ravi Fotedar
Aquatic Science Research Unit, Curtin University of Technology, Brodie Hall Building, Technology Park, Bentley, Western Australia, Australia

Research Summary

Increased concern over effects of antibiotics usage in aquaculture, on environment and human health, has prompted the search for alternative products. Recently, immunostimulants such as probiotics and prebiotics have shown promising results as preventive and environmentally friendly alternatives to the antibiotics. Among common immunostimulants used, mannan oligosaccharides (MOS) have received heightened attention in aquaculture. Since the first use of MOS in aquaculture, there has been increased number of studies demonstrating their ability to increase the survival, growth performance and control of the potential pathogens in fish and crustacean including, marron (*Cherax tenuimanus*), yabbies (*Cherax destructor*) and tropical rock lobsters (*Panulirus ornatus*). The chapter reviews the role of MOS on the culture of crayfish and fishes while detailing the effects of MOS on the growth performance, physiology and immune response of these aquatic animals. Suggestions for further research on the application of MOS in crayfish aquaculture are also included in the Chapter.

[*] Email: hmsang2000@yahoo.com.

In: Biochemistry Researcher Biographical Sketches … ISBN: 978-1-62081-308-9
Editors: J. T. Walker and C. M. Jones

Chapter 186

Facile Synthesis of Unnatural Oligosaccharides by Phosphorylase-catalyzed Enzymatic Glycosylations Using New Glycosyl Donors

Jun-ichi Kadokawa*
Graduate School of Science and Engineering, Kagoshima University, Kagoshima, Japan

Research Summary

Unnatural oligosaccharides can be expected to exhibit new functions and applications in glycoscience such as potential as drug candidates. Enzymatic glycosylation is a useful tool for the preparation of oligosaccharides with well-defined structure. α-1,4-Glucosidic linkages can be prepared by an enzymatic polymerization through the successive phosphorylase-catalyzed glucosylation using α-D-glucose 1-phosphate as a glycosyl donor. Since enzymes often express loose specificity for recognition of substrate structures, extension of the phosphorylase-catalyzed glycosylation using different substrates is useful to obtain new unnatural oligosaccharides. In this chapter, on the basis of above viewpoints, the facile synthesis of unnatural oligosaccharides by the phosphorylase-catalyzed enzymatic glycosylations using new glycosyl donors of glycose 1-phosphates is described. It has been found that α-D-xylose, α-D-mannose, 2-deoxy-α-D-glucopyranose, 3- or 4-deoxy-α-D-glucose, and α-(*N*-formyl)-D-glucosamine 1-phosphates are recognized by phosphorylase as the glycosyl donor, to occur the transfer reaction of the corresponding sugar residues to maltooligosaccharides, giving the unnatural oligosaccharides. Consequently, construction of the new oligosaccharide chains containing the different units by the phosphorylase-catalysis probably leads to development on new applications of unnatural substrates, such as the drug candidate.

* Tel: +81-99-285-7743, Fax: +81-99-285-3253,
E-mail: kadokawa@eng.kagoshima-u.ac.jp

In: Biochemistry Researcher Biographical Sketches ... ISBN: 978-1-62081-308-9
Editors: J. T. Walker and C. M. Jones

Chapter 187

Oligosaccharides: Sources, Properties and Applications

Kaoshan Chen and Yungui Bai
Laboratory of Biochemistry and Molecular Biology. School of Life Science, Shandong University, Jinan, China

Research Summary

It has been demonstrated that oligosaccharides have diverse bioactivities and functions. In plants, oligosaccharides act as early signal molecules and play an important role in the plant growth, development and morphogenesis as well as plant defense responses. Oligosaccharides are characterized as elicitors to trigger plant systemic acquired resistance against a variety of pathogens. Furthermore, oligosaccharides have been used to restore and maintain the intestinal balance by stimulating the growth and reproduction of probiotic strains such as *Lactobacillus* and *Bifidobacterium* and restraining the infection of pathogenic bacteria. In murine and human, the immunemodulatory activities of oligosaccharides which can stimulate the proliferation of splenocytes and the activation of macrophages are confirmed by many studies. Meanwhile, oligosaccharides significantly induce the accumulation of immune regulation factors and inhibit the growth of cancer cell, which act as assistant agents in treating immune system diseases such as hepatitis and cancer. Oligosaccharides also can regulate blood lipid, blood sugar and blood pressure. Therefore, these biological properties of oligosaccharides have made great contribution in the application of these oligosaccharides in agriculture, functional food and pharmaceutical products.

In: Biochemistry Researcher Biographical Sketches ...
Editors: J. T. Walker and C. M. Jones
ISBN: 978-1-62081-308-9

Chapter 188

Initiators for Anionic Polymerization: Old and New Developments

Andrés E. Ciolino, Angel J. Satti and Marcelo A. Villar
Planta Piloto de Ingeniería Química (PLAPIQUI), Departamento de Ingeniería Química, Universidad Nacional del Sur (UNS), Consejo Nacional de Investigaciones Científicas y Técnicas (CONICET). Camino "La Carrindanga", Bahía Blanca, Argentina

This chapter is dedicated to Prof. Enrique M. Vallés, for his constant support in the development of anionic polymerization techniques in Argentina.

Research Summary

Anionic polymerization techniques offer powerful methods for the controlled manipulation of macromolecular architecture, and are still today unique for the synthesis of a wide variety of *model* polymeric materials. The high–vacuum technique employed in classical anionic polymerization methodology requires highly developed and specialized skills, since the polymer scientist must possess both an adept glassblowing ability and a thorough understanding of the behavior of compounds (volatile and non volatile) in a vacuum environment. In addition, high–vacuum techniques are time consuming and usually lead to the synthesis of only a few grams of *model* polymeric materials. However, these "limitations" are a small price to pay given their potential for generating homogeneous macromolecular polymers. Complex macromolecular architectures such as linear multiblocks, branched or cyclic homo and block polymers, stars with different number of branches, μ–stars, and more complex structures, like dumbbell, model elastomeric networks or comb copolymers, can be obtained by using this powerful method if appropriate initiators and experimental conditions are chosen. In this work, we present an up-to-date review of the literature regarding initiation for anionic polymerization.

In: Biochemistry Researcher Biographical Sketches ... ISBN: 978-1-62081-308-9
Editors: J. T. Walker and C. M. Jones

Chapter 189

Advanced Organocobalt Initiators of Radical Polymerization: Current State and Potential Applications

***Ilia Levitin*[*,1], *Marina Tsarkova*[#,2] *and Inessa Gritskova*[3]**
[1]A. N. Nesmeyanov Institute of Organoelement Compounds, Moscow, Russia
[2]K. I. Skryabin Moscow State Academy of Veterinarian Medicine and Biotechnology, Academician, Moscow, Russia
[3]M. V. Lomonosov State Academy of Fine Chemical Technology, Moscow, Russia

Research Summary

The initiators in question are octahedral alkylcobalt(III) complexes with tridentate Schiff bases composed of ketoenols and chelating diamines in a 1:1 ratio. The two remaining coordination sites are occupied by either a chelating bidentate ligand or two monodentate ones. A wide range of these complexes (RCo.s) differing in reactivity, hydro-/lipo-phylity and presence or absence of functional groups transferable to polymeric products has been prepared with moderate to high yields, starting from commercial reagents.

The main feature of RCo.s is their ability to undergo ready homolytic cleavage speeded up both by acid and, though less effectively, by alkali. Unlike conventional homolytic decomposition of organic and organometallic compounds, these reactions proceed totally irreversibly at least in acidic to neutral media. The cause of such an unusual reaction behavior of RCo.s is ultimate lability of cobalt(II) complexes–primary products of the homolysis step. For this reason, RCo.s are, ceteris paribus, more effective in (co)polymerization of vinylic monomers and dienes than common chemical initiating agents. This is particularly true for

* A. N. Nesmeyanov Institute of Organoelement Compounds, Vavilov St. 28, Moscow, 119991, Russia (E-mail: ilia-levitin@mtu-net.ru)
# K. I. Skryabin Moscow State Academy of Veterinarian Medicine and Biotechnology, Academician Skryabin St. 23, Moscow, 109472, Russia (E-mail: marina.tsarkova@gmail.com)

emulsion systems. In this manner, the RCo initiators have been successfully tried in preparation of a variety of industrial polymers.

These studies have revealed substantial advantages of RCo.s over conventional, e. g. peroxide initiators. First, the use of pH-dependent organocobalt initiators provides certain improvements in conditions, effectiveness and control of radical polymerizations, which should result in economic gains as well as enhanced industrial and ecological safety. Second, characteristics of the polymer products can be upgraded and/or conveniently regulated, particularly in the case of the emulsion processes.

Certain specific fields for application of polymers and latexes produced using RCo initiators are suggested.

In: Biochemistry Researcher Biographical Sketches ...
Editors: J. T. Walker and C. M. Jones

ISBN: 978-1-62081-308-9

Chapter 190

Surface-Initiated ATRP as a Versatile Method to Hybrid Materials

Miklos Czaun[1,2], Abul K. Mallik[1], Makoto Takafuji[1] and Hirotaka Ihara[*,1]

[1]Department of Applied Chemistry and Biochemistry, Kumamoto Univeristy, Kumamoto, Japan

[2]Loker Hydrocarbon Research Institute, University of Southern California University Park Campus, Los Angeles, CA, US

Research Summary

Much research has been conducted on the development of new surface-initiated polymerization techniques since Dekking and co-workers performed the first surface-initiated radical polymerization from clay surfaces, partly driven by the increasing demand for hybrid materials. These materials have attracted increasing attention due to their potential ability to combine the properties of polymers (such as insulating behaviour or solubility in organic solvents) with the properties of inorganic materials (e.g. conductivity, and unique optical or magnetic properties). The preparation of hybrid materials includes a broad range of methods from (1) the simple physisorption of polymers; through (2) covalent immobilization of end-functionalized polymers to surface functional groups ("graft to" approach); to (3) surface-initiated polymerization from chemically attached initiators ("graft from" method). The use of the "graft from" approach has been a focus of great attention since it was recognised that higher graft densities and better control over brush thickness can be achieved due to the relatively easy diffusion of monomer molecules to the surface-attached propagating radicals.

* Department of Applied Chemistry and Biochemistry, Kumamoto Univeristy, 2-39-1 Kurokami Kumamoto, 860-8555 Japan, tel / fax: +81-96-342-3661 / +81-96-342-3662, e-mail: ihara@kumamoto-u.ac.jp.

In: Biochemistry Researcher Biographical Sketches ... ISBN: 978-1-62081-308-9
Editors: J. T. Walker and C. M. Jones

Chapter 191

Citric Acid: An Emerging Substrate for the Formation of Biopolymers

***Gurpreet Singh Dhillon*[1], *Satinder K. Brar*[*,1] *and M. Verma*[2]**
[1]INRS-ETE, Université du Québec, Québec, Canada
[2]Institut de recherche et de développement en agroenvironnement Inc. (IRDA), Québec (Québec), Canada

Research Summary

Biopolymers are conventionally defined as materials synthesized from the natural and renewable sources, such as starch, cellulose, organic acids and natural fibers. Biopolymers are a versatile class of compounds used in every aspect of life and their growth is evident by the wide spectrum of emerging applications in every sector of the economy. Biopolymers are mainly employed for industrial and biomedical applications. The industrial uses include packaging, electronics, automotive, building materials and agricultural purposes, among many others. Biopolymers in the medical sector are a new and exciting area of research and are developing at a remarkable pace with possible applications in bio-nanotechnology, medical devices, tissue engineering, and drug delivery applications. The scope of applications of biopolymers is expanding in every aspect of modern medicine today and the market share is likely to be rise in the future. Biopolymers remain a hot theme of research these days, as preference of major medical and pharmaceutical industries towards natural materials. As the biopolymers are derived from natural and renewable resources, hence possess properties, such as biocompatibility and biodegradability, which gives them an edge over conventional synthetic polymers for use in a wide range of applications. Advances in the field of biotechnology and related sciences, such as genetic engineering and recombinant DNA technology has changed the way of synthesis of biopolymers and provided highly applicable information for tailoring of different biomaterials. Today, the technology of biopolymer synthesis and fermentation process has taken a leap forward in advancement that offers

* INRS-ETE, Université du Québec, 490, Rue de la Couronne, Québec, Canada G1K 9A9, Corresponding author, Phone: 1 418 654 3116; Fax: 1 418 654 2600; E-mail: satinder.brar@ete.inrs.ca.

tremendous flexibility to design biopolymers relevant for a specific application. The number of applications of biopolymers is constantly on the rise and it is reasonable to believe that biopolymers, coupled with their unique properties and advances in production technology, would serve to bridge the huge unmet needs of the society.

In recent years, citric acid is gaining universal attention as an innocuous molecule having a broad spectrum of applications, attributed to its numerous properties. Citric acid possesses a major advantage in being biodegradable, biocompatible, non-toxic, anti-bacterial, inexpensive, easily available and highly versatile. The demand of citric acid is escalating gradually due to its generally recognized as safe (GRAS) nature. With more and more advanced application coming to light, the production of citric acid is witnessing high annual growth rate. The size of the citric acid market is huge and continues to expand, mainly because of its widespread use in food, pharmaceuticals, health-related consumer products and advance biomedicine applications such as in nano-biotechnology, tissue engineering and drug delivery. This chapter deals with the general aspects of biopolymers with special emphasis on the role of citric acid, their types, components, and finally on their wide range of applications in different industrial sectors.

In: Biochemistry Researcher Biographical Sketches … ISBN: 978-1-62081-308-9
Editors: J. T. Walker and C. M. Jones

Chapter 192

Functional Initiators in Radical Polymerization

Barbara Pabin–Szafko and Ewa Wiśniewska*
West Pomeranian University of Technology,
Szczecin Polymer Institute, Szczecin, Poland

Research Summary

Functional initiators find application in conventional radical polymerization for the synthesis of polymers with higher molecular weights, oligomers and block or graft copolymers. The initiator functionality can be considered by reason of the presence of functional end groups, such as hydroxyl and carboxyl, or azo and perester bonds, which undergo dissociation to the alkyl, alkoxy or acyloxy radicals under the influence of temperature or irradiation.

Possibilities ofor preparation of the azoester monomeric and simultaneously ditelechelic initiators by Pinner's synthesis of 2,2'-azobis(2-methylpropionitrile) (AIBN) and the homologous series of diols, from monodisperse 1,4-butanediol (1,4-BD), tri(oxyethylene)diol [Tri(OE)D] and tetra(oxyethylene)diol [Tetra(OE)D] to polydisperse oligo(oxyethylene)diols (OOED) with average molecular weight 200 and 400 g/mol were investigated. The influence of the polydisperse diol excess on a product type was determined by recognition of the mechanism for incorporating of the molecule fragment originated from polydisperse diol substrates (with different functions of molecular weight distribution).

The other type of initiators investigated were azo-peresters di-tert-butyl-6,6'-azobis-(6-cyanoperoxy heptanoate) (6,6-di-tBu) and di-tert-amyl-6,6'-azobis-(6-cyanoperoxy heptanoate) (6,6-di-tAm), synthesized on the basis of 6,6'-azobis-(6-cyanoheptanoic acid) (ACHpA).

The chemical (with determination of hydroxyl end groups) and molecular structure of initiators (IR, ^{1}H NMR spectroscopy, VPO), the thermal properties (DTA and DSC methods)

* West Pomeranian University of Technology, Szczecin Polymer Institute 10, Pułaski St., 70-322 Szczecin, Poland
Fax +48 91 449 42 47, ewa.wisniewska@zut.edu.pl.

and the kinetics of thermal decomposition of azo and perester bonds (by volumetric and gas chromatography methods) were established. The decomposition rate constants in the mixtures of monomer/solvent, in relation to monomer concentration were determined and it was found that the reaction order equaled 1 for all initiators tested.

The functional initiators were used for synthesis of ditelechelic oligomers/polymers and "active" polymers with perester groups that possess the character of macromolecular initiators, which in the presence of other monomers, allow the synthesis of block copolymers. The processes of vinyl monomers [methyl methacrylate (MMA), acrylonitrile (AN) and styrene (S)] polymerization in *N,N*-dimethylformamide (DMF) solution were investigated by dilatometric method, determining the polymerization rate and overall rate constant. The kinetic characteristics of the polymerizations were compared with model polymerization initiated with conventional initiator AIBN or initiator 4,4'-azobis-(4-cyanopentanoic acid) (ACPA) with functional end groups, carried out in similar conditions.

Azoinitiators served also the explanation of the interaction mechanisms of various radical scavengers, e.g. vitamins, and the examination of radical activity of carbonaceous nanoparticles in the form of fullerenes, nanotubes or nanofibres.

Functional azoinitiators 2,2'-azobis(2-methyl-4-hydroxybutyl propionate) (AIB-BD), 2,2'-azobis[2-methyl-ω-hydroxy-tri(oxyethylene) propionate] [AIB-Tri(OE)] and 2,2'-azobis[2-methyl-ω-hydroxyoligo(oxyethylene)propionate] [AIB-OOE] were used to prepare solvent-borne acrylic pressure-sensitive adhesives (PSA), whoseich properties were compared with the product of acrylic monomers polymerization initiated by AIBN.

In: Biochemistry Researcher Biographical Sketches ... ISBN: 978-1-62081-308-9
Editors: J. T. Walker and C. M. Jones

Chapter 193

Assessing the Effect of Selected Photoinitiators on the Properties of Photopolymerised Temperature Sensitive Hydrogels

***Luke M. Geever*[*], *John G. Lyons*[#] *and Clement L. Higginbotham*[†]**
Materials Research Institute, Athlone Institute of Technology,
Athlone, Co. Westmeath, Ireland

Research Summary

Photopolymerisation continues to be a popular method for the fabrication of biomedical hydrogels for drug delivery, cell encapsulation and tissue engineering applications. In this chapter, random copolymers based on *N,N*-diethylacrylamide (DEAAm), 1-vinyl-2-pyrrolidinone (NVP) and *N,N*-Dimethylacrylamide (DMAAm) produced by free radical photopolymerisation are discussed. To elucidate the effect of photoinitiator type on the properties of the resultant copolymers, two ultraviolet (UV) light sensitive initiators were investigated in the synthesis, namely 1-hydroxycyclohexylphenylketone (Irgacure® 184) and 2-hydroxy-1-[4-(hydroxy-ethoxy)phenyl]-2-methyl-1-propanone (Irgacure® 2959). Poly(ethylene glycol) dimethacrylate was used as the crosslinking agent. The hydrogels were synthesised to have lower critical solution temperatures (LCSTs) close to body temperature, which is favourable particularly for 'smart' drug delivery applications. The chemical structure of the xerogels was characterised by means of Fourier transform infrared spectroscopy (FTIR). Properties such as the glass transition temperature, lower critical solution temperature, pulsatile swelling and mechanical behaviour were investigated using modulated differential scanning calorimetry (MDSC), swelling analysis and oscillatory rheological

[*] Tel: 00353 90 6428054; fax: 00353 90 6424493: e-mail address: lgeever@ait.ie
[#] Tel: 00353 90 6422549; fax: 00353 90 6424493: e-mail address: slyons@ait.ie
[†] Corresponding author: Dr. Clement L. Higginbotham, Tel: 00353 90 6468050; fax: 00353 90 6424493: e-mail address: chigginbotham@ait.ie

studies. Photoinitiators commonly used in hydrogel polymerisation are discussed throughout this study, while particular attention is paid to evaluating the copolymers synthesised to establish what effect the photoinitiator chosen had on the resultant hydrogel properties.

In: Biochemistry Researcher Biographical Sketches ...
Editors: J. T. Walker and C. M. Jones

ISBN: 978-1-62081-308-9

Chapter 194

Functional Oligomeric Azo-Initiators for Grafting of Polymers on Clay Surface

Vikas Mittal*

Institute of Chemical and Bio-engineering, Department of Chemistry and Applied Biosciences, Zurich, Switzerland

Research Summary

To improve the dispersion of filler in the polymer matrices to achieve exfoliated nanocomposites, a number of different methods have been used. Grafting of polymers from the surface of the filler is an attractive approach among these methods. To achieve polymer grafting, it is required to immobilize an initiator on the surface of the filler platelets. This initiator can then be used to graft polymer brushes from the surface of the filler. Initiators for this purpose can either be monocationic or bicationic, and offer their specific limitations and advantages. The use of bicationic initiators leads to generation of less amount of free polymer than the monocationic initiator but has also higher potential of quick termination of the radicals owing to their close proximity with each other. The use of living nitroxide mediated polymerization has also been demonstrated to delay the uncontrolled termination of the radicals which leads to extensive as well as controlled grafting of polymers from the filler surface.

* Current address: BASF SE, Polymer Research, 67056 Ludwigshafen, Germany, vikas.mittal@chem.ethz.ch.

In: Biochemistry Researcher Biographical Sketches … ISBN: 978-1-62081-308-9
Editors: J. T. Walker and C. M. Jones

Chapter 195

Photoinitiating Systems Based on Unusual Radicals

Jacques Lalevée*[*], *Mohamad-Ali Tehfe, Xavier Allonas and Jean-Pierre Fouassier
Department of Photochemistry, CNRS, University of Haute Alsace, ENSCMu, France

Research Summary

Photoinitiated polymerization reactions are characterized by important advantages compared to thermal reactions. The reaction can easily be turned on or off and is intrinsically spatially controlled. Most photopolymerization reactions operate without solvent which is strongly demanded nowadays. The search for new photoinitiating systems remains important. Particularly, the design of new sources of radicals is the subject of huge research efforts. The design of new initiating radicals with different geometries and properties and/or enhanced reactivity should provide new opportunities for the development of photocurable materials. In the present paper, new photoinitiating systems generating boryl ($L\rightarrow B^{\bullet}R_2$), silyl ($R_3Si^{\bullet}$) or germyl ($R_3Ge^{\bullet}$) radicals are reviewed. These systems based on new radical chemistries improve both free radical photopolymerization (FRP) and free radical promoted cationic photopolymerization (FRPCP).

* Department of Photochemistry, CNRS, University of Haute Alsace, ENSCMu, 3 rue Alfred Werner, 68093 Mulhouse Cedex – France.Author E-mail address: j.lalevee@uha.fr.

In: Biochemistry Researcher Biographical Sketches ...
Editors: J. T. Walker and C. M. Jones

ISBN: 978-1-62081-308-9

Chapter 196

Role of Monocationic and Bicationic Initiators on the Grafting of Polymer Chains 'from' the Clay Surface

Periyayya Uthirakumar*
School of Semiconductor and Chemical Technology, Semiconductor Physics Research Center, Chonbuk National University, Jeonju, South Korea

Research Summary

Grafting of polymer chains from the clay surface are made it successful with the help of cationic azo-initiators containing quaternary ammonium functional groups. Monocationic and bicationic initiators are designed for this purpose and a simple ion-exchange method is used to prepare initiator-bound clay layers. The interlayer *d*-spacing of the resulting initiator-bound clays are largely dependent on the size and their orientation of the organic initiators. The competence between the monocationic and bicationic initiators toward the surface functionalization upon ion exchange is discussed. The chain length and diameter of initiator molecules are facilities a higher *d* spacing in bicationic initiator-bound clay than the monocationic. The successful exfoliation of polymer clay nanocomposites resulted from the polymer grafting from the clay surface makes superiority over the other techniques. Hence, they offered a very high improvement in both thermal and mechanical properties, due to the high aspect ratio of individual clay layers that are uniformly dispersed throughout the polymer matrix.

* Corresponding author: uthirakumar@gmail.com.

In: Biochemistry Researcher Biographical Sketches ...
Editors: J. T. Walker and C. M. Jones
ISBN: 978-1-62081-308-9

Chapter 197

Designing Non-Classical Inclusion Bodies

Špela Peternel [*]
Department of Biosynthesis and Biotransformation,
National Institute of Chemistry, Ljubljana, Slovenia

Research Summary

Recombinant proteins are widely used in industry (pharmaceutical, chemical, food, cosmetics...), research and diagnostics; therefore, interest in their production is increasing. Due to low cost, high productivity, simple cultivation and rapid use, bacterial systems remain very interesting for recombinant protein production. One of the most commonly used bacterial hosts is still *Escherichia coli*. However, upon overexpression in *E. coli*, recombinant proteins tend to aggregate into inclusion bodies (IBs). For a long time, IBs were considered to be deposits of biologically inactive target proteins. The denaturation and renaturation step of protein isolation from IBs have to be optimized. However, the isolation process often results in poor recovery of biologically active proteins and accounts for a major cost in protein production, thus IBs were regarded as a bottleneck in protein production.

The temperature and induction regime were recognized as very important factors affecting the growth of the bacteria *E. coli*, as well as the recombinant protein yield and protein folding and thus protein quality inside IBs. After optimization of the bioprocess, we were able to identify and explore a new sub-type of inclusion bodies. These new IBs contain correctly folded protein. They show several significantly different properties from IBs previously described (solubility, fragility, irreversible contraction at low pH); and we have therefore proposed that they be given the designation "non classical inclusion bodies" (ncIBs).

[*] Špela Peternel, Laboratory of Biosynthesis and Biotransformation, National Institute of Chemistry, Hajdrihova 19, 1000 Ljubljana, Slovenia, Telephone num.: +386 1 4760262, Fax: +386 1 4760300, e-mail: spela.peternel@ki.si.

Non-classical IBs are easily soluble in mild detergents, thus proteins can be extracted under non-denaturing conditions in biologically active form without applying any renaturation steps. Thus the isolation process is faster and economically advantageous and can be exploited for environment-friendly biotechnological production.

Such IBs composed of active proteins could also be used as pure nanoparticles in diagnostics, as biocatalysts in enzymatic processes, or even as biopharmaceuticals. Furthermore, studying IBs' formation inside prokaryotic cells could provide useful knowledge of aggregation mechanisms, applicable also to aggregates formed inside eukaryotic cells, responsible for numerous degenerative diseases.

In: Biochemistry Researcher Biographical Sketches ... ISBN: 978-1-62081-308-9
Editors: J. T. Walker and C. M. Jones

Chapter 198

Extraribosomal Functions of Ribosomal Protein S19" in Monograph Ribosomal Proteins

Tetsuro Yamamoto [*] ***and Hiroshi Nishiura***
Department of Molecular Pathology, Faculty of Medical and Pharmaceutical Sciences, Kumamoto University Graduate School, Kumamoto, Japan

Research Summary

Some ribosomal proteins play roles in activities other than translation. Such functions of ribosomal proteins are called extraribosomal function or extraribosomal activity in common. Ribosomal protein S19 (RPS19) is a component of the small subunit of eukaryotic and archaeobacterial ribosomes having 145 amino acid residues. RPS19 exhibits various extraribosomal functions, which basically correlate to the apoptosis or programmed cell death. In the apoptosis-initiated cells, RPS19 molecules are cross-linked inter-molecularly with an isopeptide bond between Lys122 and Gln137 by a transglutaminase-catalyzed reaction. This results in homo-dimerization gaining a new molecular conformation capable of binding the C5a receptor, a G protein-coupled heptahelical cell surface receptor. The dimer is then released from the apoptotic cells. Meanwhile, the apoptosis-initiated cells *de novo* synthesize the C5a receptor. The C5a receptor was initially identified as a leukocyte chemotactic receptor for the complement-derived C5a. Due to a molecular mimicry between the RPS19 dimer and C5a, both molecules can ligate the C5a receptor exhibiting some difference between them. The RPS19 dimer has a potency to cause at least three cellular events through the C5a receptor ligation. One is apoptosis promotion by an autocrine/paracrine effect. On the other hand, the RPS19 dimer recruits monocytes/ macrophages by ligating the C5a receptor on these leukocytes. The monocytes/macrophages

[*] Correspondence: Tetsuro Yamamoto, MD, Department of Molecular Pathology, Faculty of Medical and Pharmaceutical Sciences, Kumamoto University Graduate School, 1-1-1 Honjo, Kumamoto 860-8556, Japan. Email: tetsu@gpo.kumamoto-u.ac.jp Fax: +81-96-373-5308 Phone: +81-96-373-5304.

thus infiltrated engulf the apoptotic cells and digest them. The macrophages then move to regional lymph nodes *via* lymphatics and present apoptotic cell-derived antigenic peptides to T lymphocytes. The acquired immune response may then start on demand. The apoptotic cell clearance is basically an event outside the acute inflammatory response in which neutrophils play major roles. The RPS19 dimer inhibits neutrophil infiltration by ligating the neutrophil C5a receptor apparently as an antagonist. Other extraribosomal functions of the RPS19 dimer have been observed. There is a hypothesis that some elements of the apoptotic program would be used in terminal differentiation of erythroblasts when they lose their nuclei, and macrophages would be responsible for clearing the erythroblast-derived nuclei. The pro-apoptotic effect and macrophage recruiting effect of the RPS19 dimer seem to also function in the erythropoiesis. Interestingly enough, there is a molecule indistinguishable from RPS19 in plasma. The RPS19-like molecule is cross-linked by coagulation factor XIIIa, the plasma transglutaminase, in the blood coagulation process. The RPS19 dimer-like molecule seems play a role in resolution of coagula due to recruiting monocytes/macrophages.

In: Biochemistry Researcher Biographical Sketches … ISBN: 978-1-62081-308-9
Editors: J. T. Walker and C. M. Jones

Chapter 199

Order-Disorder Transitions in Ribosome Assembly

Youri Timsit[1], Frédéric Allemand[2], Claude Chiaruttini[2] and Mathias Springer[2]
[1]Information Génomique et Structurale CNRS, Parc Scientifique de Luminy, France
[2]Laboratoire de Biochimie, Institut de Biologie Physico-Chimique, CNRS*
Paris, France

Research Summary

The recent crystallographic structures of the ribosomal subunits have revealed the picture of the final product of a complex assembly process that condenses the rRNA and the ribosomal proteins into active ribosomes. The folding of rRNA is the rate-limiting step of this complicate assembly pathway. Some ribosomal proteins are required for facilitating the structural rearrangements of rRNA and avoiding the kinetic traps that frequently impede RNA folding. It is thought that the long basic r-protein extensions that penetrate deeply into the subunit cores play a key role in this process through disorder-order transitions and/or co-folding mechanisms. A current view is that such structural transitions may facilitate the proper rRNA folding. In this paper, the structures of the free and bound forms of proteins that have been experimentally found to be essential for the first steps of ribosome assembly have been compared. It is shown that the extensions of L3, L4, L20 and L22, have different structural and dynamics properties that probably correlate with different functions. This study also suggests that the specific coil-helix transition that occurs in a phylogenetically conserved cluster of basic residues of the L20 extension is strictly required for the large subunit assembly in eubacteria. In contrast, in showing that the ordering of the inner loop of L4 upon rRNA binding is not required for the assembly function of L4, this study indicates that different categories of disorder-order transitions are associated with different biological functions.

* 13, rue Pierre et Marie Curie, 75005 Paris France.

In: Biochemistry Researcher Biographical Sketches ... ISBN: 978-1-62081-308-9
Editors: J. T. Walker and C. M. Jones

Chapter 200

Disruption of Ribosomal Protein Gene Expression in Eukaryotes: A Class of Mutations Causing Similar Global Growth Defects with Cellular Variations

Catherine Kirn-Safran*

Department of Biological Sciences, University of Delaware, Newark, DE, US

Research Summary

There is much concentration in recent years in understanding the coordinated role of ribosomal protein expression in the regulation of ribosome biogenesis, translational control, and cellular functions. Although ribosomal proteins are dispensable for catalyzing the formation of peptidic bonds, they provide a conformational environment essential for the process of protein biosynthesis. Imbalance in the production of ribosomal proteins results in non-equimolar amounts of protein for each subunit and induces defects in ribosome assembly. For these reasons, delineation of the precise function of specific ribosomal protein gene products is rendered difficult as their gene targeting is accompanied by severe growth defects resulting in cellular death. The objective of this commentary is to provide a summary of ribosomal protein mutations generated in eukaryotes and discuss their proposed roles in the control of cell growth and proliferation and impact in human diseases.

* Corresponding author: Catherine Kirn-Safran, Ph.D., University of Delaware, Dept. of Biological Sciences, 310 Wolf Hall, Newark, DE 19716, USA., email ckirn@udel.edu, phone: 302-831-3249, fax: 302-831-2281.

In: Biochemistry Researcher Biographical Sketches … ISBN: 978-1-62081-308-9
Editors: J. T. Walker and C. M. Jones

Chapter 201

Ancestral Protein Reconstruction as a New Field in Protein Engineering

Ayumu Konno[1], Tomohisa Ogawa[1] and Tsuyoshi Shirai[2]
[1]Tohoku University, Sendai, Japan
[2] Nagahama Institute of Bio-Science and Technology, and
Japan Science and Technology Agency, Nagahama, Shiga, Japan

Research Summary

The extent of current research in ancient biology is not restricted by the presence of fossil records, owing to the developments in computational analysis and engineering techniques of proteins. Ancestral sequences of extant genes/proteins can be inferred with the molecular phylogeny reconstruction methods, which are mainly based on maximum-likelihood estimation. Once an ancestral sequence is determined, a reconstruction of the ancestral protein is not difficult by using the current protein engineering techniques, and reproduced protein can be experimentally tested either biochemically or biophysically. To date, such 'experimental molecular archeology' has been examined on several enzymes, elongation factor, photo-reactive proteins, nuclear receptor, and lectin, and largely helped to elucidate evolutionary processes of stability, specificity, and structure of the proteins. Although this method would promise a new field in protein engineering, several problems still remain to be solved, the largest among which should be accuracy of inferred ancestral sequence as a whole. In this chapter, the reported studies of ancestral protein reconstruction are reviewed with a focus on our recent results on the ancestral congerin, which is the β-galactoside specific lectin of fish involved in the biological defense system. The reconstruction of ancestral congerin is important because two extant isoforms of this protein, congerins I and II, have been rapidly evolving, and their molecular properties and structures have been differentiating due to the natural selection pressures operating on the genes.

In: Biochemistry Researcher Biographical Sketches … ISBN: 978-1-62081-308-9
Editors: J. T. Walker and C. M. Jones

Chapter 202

Ribosomal Proteins in Important Tropical Diseases

Viroj Wiwanitkit
Wiwanitkit House, Bangkhae, Thailand

Research Summary

The ribosome is an important cellular organelle that is necessary for the basic cellular process, protein synthesis. Ribosomal protein is a group of proteins that is presented with a focus on biomedical research. In this article, the author will review and present important data on the ribosomal proteins in important tropical diseases including malaria and tuberculosis.

In: Biochemistry Researcher Biographical Sketches … ISBN: 978-1-62081-308-9
Editors: J. T. Walker and C. M. Jones

Chapter 203

Three Challenges in Post-Proteomics Time

Shaomin Yan[1] and Guang Wu[1,2]
[1] National Engineering Research Center for Non-food Biorefinery, Guangxi Academy of Sciences, Guangxi, China
[2] Computational Mutation Project, DreamSciTech Consulting, Guangdong, China

Research Summary

In this chapter, we discuss three challenges we face in explaining the life from genome and proteome viewpoint, that is, how to measure a living protein/RNA/DNA, how to restore the missing relation cross generations and how to correlate the structure with its function. Finally, we suggest the possible solutions for these three challenges.

In: Biochemistry Researcher Biographical Sketches … ISBN: 978-1-62081-308-9
Editors: J. T. Walker and C. M. Jones

Chapter 204

Self-similar Property of Ribosomal Proteins

Chang-Yong Lee*
The Department of Industrial and Systems Engineering,
Kongju National University, Kongju, South Korea

Research Summary

Self-similar properties of the proteins in the ribosome, the ribosomal proteins, are investigated in terms of various types of fractal dimensions. We especially estimate both the chain fractal and capacity dimensions of backbones of the ribosomal proteins, together with the mass fractal dimension of the ribosomal proteins at the atomic level. The chain fractal and capacity dimensions of the ribosomal proteins are compared with that of the self-avoiding walk, which is a typical model of a polymer without interaction between monomers. These fractal dimensions of the ribosomal protein support the existence of the long and extended domain, which is hardly seen in the globular protein. We demonstrate that the mass fractal dimension of the ribosomal proteins is pertinent to explain their structural characteristics. The self-similarity also upholds the fact that the ribosomal proteins function primarily to stabilize the structure of the ribosome by both the long-extended domain of the protein penetrating into the inside of the RNA, and the globular domain interacting with the RNA on the exterior of it. These results partially, if not whole, unravel the structural characteristics of the proteins in the ribosome.

* E-mail address:clee@kongju.ac.kr.

In: Biochemistry Researcher Biographical Sketches ... ISBN: 978-1-62081-308-9
Editors: J. T. Walker and C. M. Jones

Chapter 205

Synthesizing Mammalian Cell Adhesive Surfaces using Self-Assembling Fibronectin Domains

Patricia Pereira [*,1] ***and Christopher F. van der Walle*** [#,2]
[1]Universidade de Lisboa, Lisboa, Portugal
[2]University of Strathclyde, Scotland

Research Summary

For tissue growth, organisation and repair, interaction between cells and their extracellular matrix (ECM) is critical. Fibronectin ECM renders a polyvalent ligand display for cell adhesion through binding to cell membrane integrin receptors. Mimicking this polyvalent display and designing ligands which bind specific integrin receptor subtypes will be discussed in this chapter. The central peptide motif for integrin binding is RGD (arginine-glycine-aspartic acid) but the integrin α5β1 subtype requires the synergy site on the 9th type III fibronectin domain (FIII) in addition to RGD on the 10th FIII domain. Polyvalent integrin α5β1 ligands have been synthesised using a protein engineering approach, designing chimeras consisting of a N-terminal 9th-10th FIII domain pair, IgG-derived hinge and leucine zipper-derived helix; the latter mutated to yield di-, tri- and tetrameric coiled coils and thus self-assembling, multimeric integrin α5β1 ligands. A unique C-terminal cysteine was appended to the helix to facilitate 'anchoring' of the chimeras with a defined orientation on a surface. Careful characterization of such self-assembling chimeras is required and can be achieved through the use of complementary techniques including size exclusion chromatography, analytical ultracentrifugation and circular dichroism. Surface adsorption and specific immobilization can then be unambiguously characterized by neutron reflectivity, relating the

* Patricia Pereira, PhD, iMed.UL - Research Institute for Medicines and Pharmaceutical Sciences, Faculdade de Farmácia, Edifício E, Universidade de Lisboa, Av. Prof. Gama Pinto, 1649-003 Lisboa, Portugal. Tel: +351-217946400 (Ext: 245), Fax: +351-217946470.
[#]C. F. van der Walle, PhD, Strathclyde Institute of Pharmacy and Biomedical Sciences, University of Strathclyde, 27 Taylor St, Glasgow, G4 0NR, Scotland. Tel: +44 (0) 141 548 5755, Fax: +44 (0) 141 552 2562.

determined models of the adsorbed protein layers to their biological function. Application of these engineered proteins and cell adhesive surfaces as scaffolds or other biomedical devices has the potential to promote tissue repair and regeneration, for a wide variety of tissues including bone and skin. These applications will be discussed in the context of the future outlook for these ECM protein engineering approaches.

In: Biochemistry Researcher Biographical Sketches ... ISBN: 978-1-62081-308-9
Editors: J. T. Walker and C. M. Jones

Chapter 206

Halophilic Adaptations of Proteins*

Albert Bolhuis*†, *Daniel Kwan and Judith R. Thomas
Department of Biological Sciences, University of Warwick, UK

Research Summary

Halophilic organisms thrive in a wide variety of environments, ranging from mildly saline to saturated brine. Extremely saline conditions are very harsh to proteins and can result in problems such as loss of flexibility, which leads to loss of catalytic activity, or aggregation and precipitation due to salting-out effects. Halophilic proteins have therefore acquired specific adaptations to prevent such problems, and that is clearly reflected in their amino acid composition. As a consequence, halophilic proteins do not function properly in low salt conditions, and many even unfold and denature without sufficient amounts of salt. One of the most obvious adaptations found in halophilic proteins is an excess of acidic residues, while other adaptations include a low lysine content and an increase in small hydrophobic amino acids, the latter of which is at the expense of large hydrophobic residues. Different functions have been suggested for the role of the acidic residues. They are, most likely, important for binding of essential water molecules, but may play also a critical role in binding of salt or prevention of aggregation. The other changes in amino acid composition in halophilic proteins are probably important for the reduction of hydrophobicity, in order to prevent aggregation and maintain flexibility in high salt concentrations. Structural features that may be important for halophilic adaptation are also found, such as an increased number of salt-bridges. That is, however, not used by all halophilic proteins, indicating that different strategies have evolved to cope with very saline conditions.

---

* A version of this chapter was also published in Protein Adaptation in Extremophiles, edited by K. S. Siddiqui and T. Thomas published by Nova Science Publishers, Inc. It was submitted for appropriate modifications in an effort to encourage wider dissemination of research.

† Corresponding author. E-mail: a.bolhuis@bath.ac.uk. Current address: Department of Pharmacy and Pharmacology, University of Bath, Bath BA2 7AY, UK.

In: Biochemistry Researcher Biographical Sketches ...
Editors: J. T. Walker and C. M. Jones
ISBN: 978-1-62081-308-9

Chapter 207

Alkaline Adaptation of Proteins*

Tsuyoshi Shirai[1,†], Tohru Kobayashi[2], Susumu Ito[3] and Koki Horikoshi[2]

[1] Department of Bioscience, Nagahama Institute of Bio-science and Technology, and JST-BIRD, Nagahama, Japan

[2] Japan Agency for Marine-Earth Science and Technology (JAMSTEC), Yokosuka, Japan

[3] Creative Research Initiative (CRIS), Hokkaido University, Sapporo, Japan

Research Summary

Alkaliphilic microorganisms (alkaliphiles) produce a variety of proteins, mostly enzymes, that are stable and optimally active at alkaline pH. The alkaline enzymes represent most of the currently known proteins adapted to high pH. Alkaline enzymes are widely used for industrial or commercial purposes, for example, as additives to laundry detergents. They have been improved in the course of evolution to be resistant to low proton density conditions that would cause protein denaturation or dissociation. In order to elucidate the mechanism of high-pH adaptation, the alkaline enzymes have been characterized and their structures have been compared with homologous non-alkaliphilic proteins. These comparative studies, together with several experimental and theoretical studies, have suggested three major strategies for alkaline adaptation: p*K*a modulation, Asp+Glu gain, and Asp+Lys loss–Glu+Arg gain. The history of representative alkaline enzymes, proposed strategies for high-pH adaptation, and protein engineering studies performed in order to adapt proteins to alkaline conditions will be reviewed in this chapter.

---

* A version of this chapter was also published in Protein Adaptation in Extremophiles, edited by K. S. Siddiqui and T. Thomas published by Nova Science Publishers, Inc. It was submitted for appropriate modifications in an effort to encourage wider dissemination of research.

† Corresponding author. E-mail: t_shirai@nagahama-i-bio.ac.jp.

In: Biochemistry Researcher Biographical Sketches ...
Editors: J. T. Walker and C. M. Jones
ISBN: 978-1-62081-308-9

Chapter 208

Preface to the Inaugural Issue of the Journal of Biochemistry and Molecular Biology in the Post-Genomic Era

Mary M. Y. Waye
School of Biomedical Sciences,
The Chinese University of Hong Kong, Shatin, N.T. Hong Kong

Research Summary

Several years ago, when I was first asked to become the president of the Hong Kong Society of Biochemistry and Molecular Biology, I toyed with the idea of establishing an official journal for the society simply because I felt that every respectable society has a journal affiliated with it. However, teaching and family commitments were very demanding, thus preventing the journal's appearance.

A couple years ago, the thought was re-ignited during a meeting with Prof. Tom Boyd, one of the founding members of the Hong Kong Society of Biochemistry or Hong Kong Biochemistry Association, the predecessor society of the Hong Kong Society of Biochemistry and Molecular Biology. I was intrigued to learn that he had tried, against all odds, to establish a journal with the name "Contributions of the Hong Kong Biochemistry Association" and he showed me with much pride one of the articles he published in one of the two issues of that earlier attempt. During a visit to Oxford University, I also had the chance to get some words of wisdom from Prof. Anthony Watts, the managing director of the European Biophysics Journal, who advised me that I really need to expand the contributors and audience of such a journal beyond Hong Kong or South East Asia to make it viable.

The real breakthrough came when there was a reorganization of our department in June 2009, which meant a bit fewer teaching duties and more time for other adventures, at least at the beginning after we split with our twin Biochemistry Department that belong to the Faculty of Science, and merged with the other pre-clinical departments to form the new School of

Biomedical Sciences under the Faculty of Medicine. After adapting to the newly formed School of Biomedical Sciences and retraining my new technician, I could not resist the temptation to establish a new journal anymore, particularly when the opportunity unfolded as Nova Science Publishers were also looking to expand their journal selections.

With this inaugural issue of a new journal, I wish to capture an audience who will be amazed and bewildered by the power of a general biochemistry and molecular biology journal. As I start out on this project, I really do not know where it will lead, but I feel strongly that there are many stories and messages that we could help transmit that will not be provided by other journals. It is a source of considerable gratification that friends have responded well to the news of the birth of this new journal and indeed I feel privileged to be in the both challenging and humbling position to try to provide something that is of value to so many colleagues and friends. However, I suspect that as our technology in communications evolves, the older method of journal publication might be somewhat outmoded and multiple strategies need to be developed to communicate our points of view as the scientific literature transforms at ever greater rates.

Some might question the journal title's length and ask whether we did reach the "Post-Genomic Era" or not. As a matter of fact, I once heard that Professor Lap-Chee Tsui seems to prefer the idea that we are actually "In the Genomic Era" rather than "Past the Genomic Era" as the title of this journal might suggest. Whatever cutoff line we wish to set, or era we wish to embrace, I feel that we are now beginning an era in which we already know much about a multitude of genomes of different species. We might not know or appreciate everything that the different genomes reveal, but we have overcome some of the earlier resistance to the use of the latest sequencing technologies, and many scientists have learnt how to do biological research using the new power of careful analyses of the genomic information. Early resistance to obtain genomic information from some scientists is not unique in the history of science, as both da Vinci and Galileo encountered steep resistance to their use of the newest technologies in their own era.

For a journal that might been seen as too broad by some, especially to those who do not read journals unless they are in their own specialties, I have chosen to provide an introduction that helps to build a case for a need of a journal with such a diverse scope. I believe that a journal of this nature can survive because in an era when websites on the Internet have rapidly become the preferred way of data mining, the gap between publishing in a broad-based journal and a more focused journal has somewhat narrowed. The habit of reading many articles from one specialized journal from the first page to the last page has given way to a few key strokes by a search engine, so that the chance of something being buried in a broad topic journal is lessened. Many scientists have been trained to do research in a limited area and pursue their topic exhaustively. In this journal, I wish to communicate a larger scope, hoping that readers will enjoy thinking further outside the comfort zone of their specialty and in so doing, achieve a wider perspective. Hopefully some will find a fresh new way of looking at their own project from a greater height.

Well meaning friends have also questioned the justification of submitting articles in a new journal with no impact factor initially but risking impressing their superiors. Over the years, conventional academic measures and practices have been linked to the ability to publish in high impact factor journals, preferably well-established ones with wide circulation. Nevertheless, we should keep in mind that in this new era, the technologies move so quickly that some work becomes obsolete after we wasted much time in answering questions from referees from a few extremely competitive journals. As many of us could hardly meet the various deadlines for grant applications or grading students' examination scripts, there is a balance between how much time one wishes to invest in trying different competitive journals and to spend time in actually doing the research that will give gratifying novel information that can also have genuine impact on one's research topic. If one takes Einstein's example alone, there is no reason other than convention, really, to link impact factor and citation counts with novelty or great accomplishments.

I hope that this journal will transmit the basic ideas in ordinary language without flowery elaborations. I also hope that it will continue to appeal to a diverse general audience, including those who are not particularly fond of reading journals of biochemistry and molecular biology with a range of broad topics. With this new journal, we might realize the potential for the cross over from different fields that have long been largely closed to specific groups of scientists. Lastly and importantly, I hope this journal will serve as a platform for young scientists to envision the solutions to difficult problems in biochemistry and molecular biology.

Mary M.Y. Waye, Ph.D.
Editor-in-chief, Journal of Biochemistry and Molecular Biology in the Post Genomic Era
President, Hong Kong Society of Biochemistry and Molecular Biology
Founding Director, Croucher Laboratory for Human Genomics
Professor, School of Biomedical Sciences, The Chinese University of Hong Kong
Shatin, N.T., Hong Kong

September, 2010

In: Biochemistry Researcher Biographical Sketches … ISBN: 978-1-62081-308-9
Editors: J. T. Walker and C. M. Jones

Chapter 209

Biochemistry and Molecular Biology in Hong Kong: Past and Future

T. R. C. Boyde*
Emeritus Professor of Biochemistry,
University of Hong Kong

Research Summary

When Frederick Gowland Hopkins attended the City of London School, seventy years before me, the word biochemistry did not exist in English. Even when he was invited to Cambridge by Michael Foster it was to teach "the chemical side of physiology" though within a few years Biochemistry there had broken away from its parent, leading the way for many departments like it because chemically-minded people wished to shed their physiological strait-jacket. Cambridge took the plunge in 1914, the University of Hong Kong (HKU) not until 1965 and Professor Walshe arrived the following year. At Chinese University of Hong Kong (CUHK) the alternate route was followed. Since no medical school yet existed Biochemistry began in science faculty, and by 1973 the department under Professor Ma Lin was far better developed than HKU. They had all the things we lacked, though there was no clear route to research funding for CUHK people or anyone else in Hong Kong. There was no Croucher Foundation, no research grant committees (such as GRF funds from the University Grants Council or UGC-8, ITF funds from the Innovation and Technology Commission or ITC- 9, or RFCID/HHRSF funds from the Food and Health Bureau-10), nothing. It all seems like a century ago.

Ted Walshe's years at HKU were frustrating in many ways. He started a program in science faculty, generating resentment which still endures among those who felt that they were losing students rather than gaining a partner; and was once heard to say, "If they want a — mess they can have a — mess", where 'they' means the Faculty of Medicine: so bad treatment by Faculties was reassuringly even-handed. To be fair, times were still tough in

* E-mail: tom@boyde.com.

Hong Kong; no-one wanted to invest in basic science and it showed. On my first visit to the HKU biochemistry department, in 1972 after Ted's early death, the words that sprang to mind were, "marble halls": there was no-one at the benches and nothing going on. One room designated as the "Research Laboratory" was the demonstrators' domain; there was a cold room and an ultracentrifuge but scarcely any other functional modern equipment, the best spectrophotometer having been damaged some time earlier by an acid spill and never repaired; no bench for me to use until a new one was built. Typical of those bad old days, a second ultracentrifuge lay hidden elsewhere in the Li Shu Fan building; bricked up when the lecturer who owned it left the place and rediscovered twenty years later. (We did put some of the rotors into use but the drive was obsolete and beyond repair.)

That's the raw truth about Biochemistry in Hong Kong, just over 30 years ago. It's hard to imagine now with all the sophisticated new buildings and advanced research in progress everywhere, but then it was a desperate matter of survival to raise the profile of the science, stimulate interest generally in doing research or paying for it to be done, and provide for cooperation and communication between the perfectly sound biochemists who were already in post. That is why the old Hong Kong Biochemical Association (now renamed as the Hong Kong Society of Biochemistry and Molecular Biology -12) came into existence, the final spark coming from Takashi Murachi of Kyoto during a flying visit in 1977, "Tom, maybe you're right not to push, but there's nothing wrong with giving the flywheel a good spin."

So we did. The Association held its first conference, entitled "Science for a Better Life", in December 1978. It was essentially a propaganda exercise directed at the general public. 200 people attended lectures in the old Chemistry Theatre of HKU and even today the papers look incredibly up-to-date and relevant. We had Lester Packer talking about bio-mass as a fuel source (carbon footprint before it became popular), Brian Morton on conservation of the sea shore, and so on. Winding up, I called attention to the need for new approaches to research funding in Hong Kong, "To achieve this .... requires a reserve fund, independent of government or of the individual researcher himself; in other words a Research Foundation. No such body exists in Hong Kong today, in one of the richest countries of the world." (Contributions of the Hong Kong Biochemical Association, vol 1, no.1, page 77).

It took ten more years to get there and by roundabout ways. The Shirley Boyde Trust made its first research grants in 1989, quickly followed by the Croucher Foundation, but still barely more than 20 years ago.

Those stories abut how biochemistry came to Hong Kong are all very well but maybe the future is more interesting, and to talk about that depends on understanding what biochemistry and molecular biology actually are. The latter phrase is used by many people to indicate that what they do is exciting biology and not dull old chemistry; again it seems to them an escape from a strait-jacket; biochemists were the little grey men obstructing progress. But the very words 'molecular biology' had been used for twenty years before the discoveries of Watson and Crick, to mean biochemistry and nothing else. It does not belittle the great contributions of those who wish to call themselves molecular biologists if we gently point out that means to sequence polypeptides and nucleic acids were both invented by a Cambridge biochemist of the old school, and similarly with most of the other fundamental advances.

Peter Campbell was a good friend of ours and he thought that biochemistry as an independent subject was no longer necessary, its job was done. This is to understand biochemistry as a servant of biology, which many biologists would like it to be, and as a finite body of knowledge which perhaps says more about Campbell as a research worker than anything else. Ted Walshe had a clearer idea of the subject; his concern was with 'the chemical phenomenon that we call Life'. My only quibble there is that he has defined life as a chemical phenomenon, an axiom rather than the thing required to be demonstrated. We might find it by experiment to be something else!

One thing that causes a lot of trouble is to equate biochemistry with biological chemistry, i.e. the study of biological compounds as objects of chemical interest. 'Organic chemistry' originally meant the chemistry of living organisms so it sounds like the same thing as biochemistry, but the phrase was then extended to all compounds of carbon. Today, there is hardly any resemblance between the work of an organic chemist and a biochemist, and when a biochemist does some synthetic work of his own he is acting as an organic chemist, a displaced person.

Suddenly everything changes when it comes to structural and physical studies of molecules having biological interest; so a better definition of biochemistry might be 'the application of chemical disciplines to the study of living organisms'. We should be thinking chemically but in a biological environment. Included must be any kind of chemistry and perhaps physical aspects are more important than the others.

Is that just playing with words? We who are true biochemists don't think so and we may wish to exclude those who apply chemical methods but not chemical thinking to their research. To determine the nucleic acid sequence of an organism is, nowadays, either biology or chemistry, not biochemistry. Our interest begins when we wish to find out how a segment of nucleic acid is prepared for action in gene expression; why one species differs from another; why so much of human DNA is junk and does nothing (– a likely story, by the bye); or the cause of cancer.

It seems to me that biochemistry/molecular biology will continue to exist as an independent discipline for as long as science itself and be defined by two things; the nature of the subject matter and the basis of the approach to studying it; not chemical methods only, nor only applied objectives, but Life as a proper subject and chemical thinking as allowing an approach peculiarly its own. And Hong Kong may have a very important part to play in that: we oldsters certainly struggled to make it possible and hope that the effort was not in vain.

In: Biochemistry Researcher Biographical Sketches … ISBN: 978-1-62081-308-9
Editors: J. T. Walker and C. M. Jones

Chapter 210

Calpains-Mediated PTEN Cleavage in Excitotoxic Neuronal Death

***Minghui Jessica Chen*[*,#,1], *Piang Chin Benjamin Soo*[*2], *Meng Shyan Choy*[3], *Zhao Feng Peng*[4], *Heung-Chin Cheng*[2] and *Nam Sang Cheung*[#,*,1,5]**

[1]Menzies Research Institute School of Medicine, University of Tasmania, Hobart, Tasmania, Australia

[2]Department of Biochemistry and Molecular Biology, Bio21 Institute of Molecular Science and Biotechnology, University of Melbourne, Parkville, Victoria, Australia

[3]Department of Biochemistry, Yong Loo Lin School of Medicine, Faculty of Science, National University of Singapore, Singapore

[4]Key Laboratory of Biogeology and Environmental Geology of the Ministry of Education, China University of Geosciences, Wuhan, P.R. China

[5]School of Life and Environmental Sciences, Deakin University

Research Summary

The tumor suppressor PTEN (phosphatase and tensin homolog deleted from chromosome 10) has been implicated in the pathogenesis of numerous neurodegenerative diseases, and is particularly noted for its inhibition of the pro-survival PI3K-Akt pathway. Excitotoxicity due to over-stimulation of ionotropic glutamate receptors in neurons are known to contribute to neurodegenerative diseases such as Alzheimer's disease and ischemic stroke. Previously we have demonstrated that the proteasomal inhibitor, lactacystin, induced cleavage of PTEN to a truncated 50kDa fragment which accumulates to the detergent-insoluble membrane fraction of the neuronal cells. Herein, we report that overstimulation of the ionotropic glutamate

[*] Minghui Jessica Chen and Piang Chin Benjamin Soo contributed equally to this study.

[#] Corresponding author and to whom proofs and reprint requests should be addressed: A/Prof Steve Nam Sang Cheung, Menzies Research Institute, School of Medicine Private Bag 23, University of Tasmania, Hobart, Tasmania 7001 Australia. Tel: +613 62262710 Fax: +613 6226 7704 E-mail: nscheung@utas.edu.au.

receptors with excess NMDA while causing induction of calpains activation in cultured cortical neurons results in the formation of a novel truncated PTEN fragment of molecular weight <50kDa. *In vitro* cleavage assay further confirmed that calpains-mediated cleavage is responsible for the expression of <50kDa, while the 50kDa PTEN fragment observed in latacystin-mediated neuronal death is held accountable by caspase-3 activation. Both truncated PTEN fragments preferentially partition to the detergent-insoluble fraction of neuronal cells. The fragments did not react with the phospho-specific PTEN antibody targeting the C-terminal regulatory serine and threonine residues, suggesting that the cleavage sites are localised at the C-terminal tail or the cleavage enhances dephosphorylation of these C-terminal regulatory sites. Future investigation to define the function and regulation of calpain- and caspase 3-generated truncated PTEN fragment will shed light on the role of PTEN in proteasome inhibition- and excitotoxicity-induced neuronal death in neurological diseases.

In: Biochemistry Researcher Biographical Sketches ... ISBN: 978-1-62081-308-9
Editors: J. T. Walker and C. M. Jones

Chapter 211

NUDCD2 Is Overexpressed in Head and Neck Squamous Cell Carcinoma and Is Involved in Nuclear Division*#

Anne Cromer[1], Alberto Zambrano[1], Régine Millon[2], Joseph Abecassis[2] and Bohdan Wasylyk[1,#,*]
[1]Institut de Génétique et de Biologie Moléculaire et Cellulaire, CNRS/INSERM/ULP, France
[2]UPRES EA 34-30, Centre Paul Strauss, Strasbourg, France

Research Summary

Head and Neck squamous cell carcinoma (HNSCC) is the sixth most common cancer in men and has a poor prognosis. The molecular events leading to HNSCC need to be understood. We have identified by Differential Display many novel deregulated genes with unknown functions that require characterization. We show here that one of these genes, *NUDCD2*, is overexpressed in 70% of patients with HNSCC and in many HNSCC cell lines in culture. *NUDCD2* RNA and protein localize to tumour cell islets of human HNSCC compared to surrounding stromal tissue. NUDCD2 overexpression in cells alters cell and nuclear shapes, and induces multinucleation. Time-lapse microscopy showed that there are mitotic aberrations and retention of nucleoplasmic bridges (NPB) between un-separated nuclei following abnormal mitosis. NUDCD2 possesses a p23-like domain, linking NUDCD2 to a family of proteins related to p23, a co-chaperone for Hsp90. Interestingly, the NUDCD2 C-terminal region is also highly similar to the Nuclear Move domain of NudC, a protein

* Financial support: (a) the Ligue Nationale Française contre le Cancer, the Ligue Régionale contre le Cancer (Haut-Rhin et Bas-Rhin), the Centre National de la Recherche Scientifique, the Institut National de la Santé et de la Recherche Médicale, the Association pour la Recherche sur le Cancer, Procore Project N° 22010TL; (b) the Ministère de la Recherche et Technologie and the Association pour la Recherche sur le Cancer for fellowships for CA.
# Requests for reprints: Bohdan Wasylyk, IGBMC, 1 rue Laurent Fries, BP 10142, 67404 Illkirch , France. Tel: + 33 3 88 65 34 11. Fax: + 33 3 88 65 32 01. Email: boh@igbmc.fr.

involved in the nuclear migration cycle and whose overexpression provokes multinucleation. Furthermore, we found that both proteins colocalize in the Golgi apparatus. These findings suggest that NUDCD2 is a new member of the NudC sub-family that plays an important role in the reformation and maintenance of nuclear integrity following mitosis. Imbalances in the levels of NUDCD2 could be evaluated as a factor contributing to chromosomal instability in HNSCC through effects on the complex processes of mitosis and cell division.

In: Biochemistry Researcher Biographical Sketches ...
Editors: J. T. Walker and C. M. Jones

ISBN: 978-1-62081-308-9

Chapter 212

Genotype Analyses Using SNP (Using MALDI-TOF Mass Spectrometry) and STR (Microsatellite) Markers in the Determination of Zygosity Status of Chinese Twins

Cadmon K. B. Lim[1], Venus S. Y. Yeung[1], Tze Lun Yeung[1], Aaron C. Y. Tam[1], Connie S. H. Ho[2], Simpson W. L. Wong[2], Bonnie W. Y. Chow[2], Yee Miao Ho[2], Crystal Hin Ning Chou[2], Chi-Hang Fred Fung[1], Chee Fei Wong[1] and Mary M. Y. Waye*[1]

[1]Croucher Laboratory for Human Genomics, School of Biomedical Sciences, The Chinese University of Hong Kong, Shatin, N.T. Hong Kong

[2]Department of Psychiatry, The University of Hong Kong

Research Summary

It has been argued that single nucleotide polymorphisms (SNPs) is a very useful tool complementary to microsatellite analyses in the near future due to its relative ease of automation and interpretation. However, because allele frequencies can vary greatly among populations, it is important to validate and identify the informative SNPs that are useful for sample identification in biomedical and epidemiological studies. We have genotyped 768 individuals (384 pairs of twins) with a panel of SNPs based mostly on the SNPs chosen by an earlier work conducted by Lee et al., 2005 for Koreans, which we hypothesized to be useful for our Chinese subjects. The MALDI-TOF mass spectrometry based iPLEX Gold assay on the MassARRAY® Platform method using Sequenom was used for determination of the

* Corresponding author: ary M.Y. Waye, Ph.D. Director, Croucher Laboratory for Human Genomics, Rm 608 MMW Bldg., School of Biomedical Sciences, The Chinese University of Hong Kong, Shatin, N.T. Hong Kong; Tel: (852)2609-6874; Fax: (852)2603-7732; Email: mary-waye@cuhk.edu.hk.

zygosity status of the twins. The ABI microsatellite kit (AmpFISTR identifiler PCR amplifcation kit) for measuring 16 loci was used to confirm the zygosity status. We found that some of the SNPs (2/22) used in Lee's paper were not suitable for our Hong Kong Chinese samples. The zygosity status as determined by the mass spectrometry method can be validated with microsatellite method using 76 samples. In summary, we have studied a panel of 25 SNPs on their effectiveness in distinguishing identical and non-identical twins (768 individuals) for the Hong Kong Chinese population using the iPLEX method of Sequenom. It is important to optimize SNP panels for genotyping depending on the population and the platform used for the studies.

In: Biochemistry Researcher Biographical Sketches ... ISBN: 978-1-62081-308-9
Editors: J. T. Walker and C. M. Jones

Chapter 213

The Dyslexia Candidate Gene Kiaa0319L Encodes N-Glycosylated Isoforms That Form Homo-Dimers

Ming-Wai Poon, Hoi-Ling Chan, King-Poo Lim and Mary Miu-Yee Waye *

School of Biochemical Sciences, Laboratory for Human Genomics,
The Chinese University of Hong Kong, Shatin, N.T., Hong Kong, China

Research Summary

KIAA0319-Like (*KIAA0319L*) *gene* has been reported to be associated with developmental dyslexia. It has been suggested to be a functional and positional candidate for developmental dyslexia, as it is a homologue of *KIAA0319* and positioned near the developmental dyslexia susceptibility locus, DYX8. This article documents the expression pattern of this gene in various human tissues, including the adult and fetal brain, by quantitative reverse-transcriptase polymerase chain reaction (RT-PCR). *KIAA0319L* mRNA was present in both adult and fetal brain, and was found to be most abundant in the cerebellum.

Kiaa0319L protein was shown to be mainly localized in the cytoplasm, the periphery of the nucleus and also within the nucleus. Experiments with overexpressed constructs containing the Kiaa0319L protein revealed that the full-length Kiaa0319L protein is an N-glycosylated protein which exists in at least 2 glyco-isoforms. This protein was able to bind to itself in a co-immunoprecipitation experiment. These properties of Kiaa0319L protein are also similar to its homologue, the Kiaa0319 protein, which is the N- and O- glycosylated plasma membrane isoform A studied by an Italian group.

* Corresponding author: Prof. Mary Waye, School of Biomedical Sciences, Croucher Laboratory for Human Genomics, The Chinese University of Hong Kong, Shatin, N.T., Hong Kong, China. Tel:852-26096874 Fax:852-26037732; E-mail: mary-waye@cuhk.edu.hk

We report here a novel study comparing the two Kiaa proteins, their protein properties and predicted domain structures. This is an important initial step to the functional studies of the two Kiaa proteins.

In: Biochemistry Researcher Biographical Sketches … ISBN: 978-1-62081-308-9
Editors: J. T. Walker and C. M. Jones

Chapter 214

Identification of Feed Forward Loops Composed of MicroRNAs and Transcription Factors in Arabidopsis

Nicholas Wu, Xiaozeng Yang and Lei Li*
Department of Biology, University of Virginia,
Charlottesville, VA, US

Research Summary

In eukaryotes, differential gene expression is regulated at many levels and by various trans-acting factors in the context of gene networks. Transcription factors (TFs) and microRNAs (miRNAs) are two classes of primary regulators for differential gene expression during organism development and function. Transcriptional gene networks mediated by TFs have been intensively mapped in several model organisms. However, relatively little has been done to examine regulatory network involving both miRNAs and TFs. This paper attempts to discover computationally the composite feed forward loops (FFLs) involving both miRNAs and TFs in the model plant *Arabidopsis thaliana* at the genome scale. 99 positional weight matrixes (PWM) that represent different DNA protein-binding motifs and 176 miRNAs were used to infer regulatory interactions at both the transcriptional and post-transcriptional levels. VisANT was then used to integrate the data and perform network motif searching. This approach generated many potential FFLs that are preferentially associated with conserved miRNAs. Further validation and study of these FFLs should provide much needed insights into the function and evolution of miRNA-mediated gene circuits in plants.

* Corresponding Author: Lei Li, Affiliation: Department of Biology, University of Virginia, Charlottesville, VA 22904-4328, USA. Phone: 1-434-982-5481. Fax: 1-434-982-5626. Email: ll4jn@virginia.edu.

In: Biochemistry Researcher Biographical Sketches ... ISBN: 978-1-62081-308-9
Editors: J. T. Walker and C. M. Jones

Chapter 215

Sanfilippo Syndrome: Identification and Removal of a Cryptic Splice Site in Alpha-N-Acetylglucosaminidase cDNA for Potential Improved Expression and Enzyme Replacement Therapy

Sarah N. Truelson, R. Rebecca Jantzen and Francis Y. M. Choy*
Department of Biology, Biomedical Research Centre, University of Victoria
Victoria, British Columbia, Canada

Research Summary

α-N-Acetylglucosaminidase (Naglu) is a lysosomal acid hydrolase that is deficient in patients with Mucopolysaccharidosis III type B (MPS IIIB). Naglu deficiency causes cytotoxic accumulation of heparan sulfate in lysosomes; this is especially detrimental to the central nervous system, resulting in neurodegeneration, behavioural problems, mental retardation, and shortened lifespan. Our laboratory is currently working to express Naglu fused to a synthetic protein transduction domain with the ultimate goal of developing an efficient system to produce human recombinant Naglu for enzyme replacement therapy. Conjugation of Naglu to a protein transduction domain may allow the enzyme to cross the blood-brain barrier to treat the neurological symptoms of MPS IIIB.

The Naglu fusion protein has been successfully produced in an active state in recombinant *Spodoptera frugiperda* 9 (*Sf*9) cultures; however, low resultant expression levels have prevented further characterization studies. Degradation of Naglu mRNA due to activation of a cryptic splice site in human Naglu cDNA was hypothesized, as we consistently

* Corresponding author: F.Y.M. Choy, Department of Biology, Biomedical Research Centre, University of Victoria, 3800, Finnerty Road, Victoria, BC, Canada V8P 5C2. Tel. (250)721-7107; FAX: (250)721-7120; E-mail: fchoy@uvic.ca.

detected the presence of a small, non-target DNA band in reverse transcription polymerase chain reaction (RT-PCR) experiments. Sequence analysis showed that the 426 bp band contained Naglu sequence corresponding to the 5' and 3' ends of the gene, with a large central portion missing, and was likely the result of cryptic splicing in the *Sf9* expression system. Site-directed mutagenesis by fusion PCR was used to alter the splice donor and acceptor sites in Naglu cDNA; RT-PCR on RNA isolated from *Sf9* cells transfected with the new construct indicated presence of the full length 2.2 Kb Naglu product and absence of the 426 bp band. Use of this construct, which successfully eliminated cryptic splicing, could potentially increase recombinant Naglu expression in *Sf9* cells.

In: Biochemistry Researcher Biographical Sketches ... ISBN: 978-1-62081-308-9
Editors: J. T. Walker and C. M. Jones

Chapter 216

Conference Report of 2010 Inter-University Post-Graduate Symposium

Edwin Chan
Department of Biochemistry, The Chinese University of Hong Kong,
Shatin, N.T. Hong Kong

Research Summary

The Department of Biochemistry, The Chinese University of Hong Kong hosted the first Inter-University Biochemistry Postgraduate Symposium on 15 May, 2010. Generously sponsored by The University Grants Committee of Hong Kong and Chung Chi College, The Chinese University of Hong Kong, this event was jointly organized with sister departments at HKU and HKUST with an aim to promote inter-institutional and inter-disciplinary communications among postgraduate students and faculty members. Over 180 postgraduate students and faculty members attended. To mark the opening of the symposium, Prof. Zihe RAO, President of Nankai University and a leading figure in the field of protein structural biology, delivered a featured lecture on "Structure-based anti-viral research". More than 70 students presented their research findings at this inaugural symposium.

Further details can be found: http://www.bch.cuhk.edu.hk/bch2010/index.html

In: Biochemistry Researcher Biographical Sketches ... ISBN: 978-1-62081-308-9
Editors: J. T. Walker and C. M. Jones

Chapter 217

Launching Theme-Based Research in Hong Kong

Anthony Chan
Research Grants Council of Hong Kong

Research Summary

An HK$18 billion Research Endowment Fund was established with allocation from the Hong Kong Government. HK$4 billion of the Fund is set aside for Theme-based Research which will support projects on themes of a longer-term nature and strategically beneficial to the development of Hong Kong. A Steering Committee, with membership consisting of the academia, representatives from different sectors of the community and the Government, and the Chairman of Research Grants Council (RGC) of Hong Kong, was set up with the major charge to identify the themes.

To launch the Scheme, funding will be focused initially on a small number of themes most beneficial to Hong Kong's strategic development and where the local academia have a competitive advantage, says RGC Chairman Professor Roland Chin.

Professor Chin points out that the research themes should aim to find solutions to important problems that are relevant to Hong Kong. The Theme-based Research Scheme's Principal Investigators must be academics from University Grants Committee (UGC)-funded institutions. It is hoped that the initiative will galvanize our strength and interest to work on high impact cutting edge research projects, and at the same time create state-of-the-art teaching materials and educational programs, as well as provide quality training for our research students.

Professor Chin says that the chosen themes are "Promoting good health", "Developing a sustainable environment" and "Strengthening Hong Kong's strategic position as a regional and international business centre".

Projects under the theme "Promoting good health" should aim to make Hong Kong a healthier society whose capability to prevent epidemics will set international standards. Research can focus on the fundamental life sciences, disease prevention and control, Chinese medicine, healthy living, aging, and health care policy, etc. Moreover, the Government's strategy of promoting medical services as one of the six priority economic areas is in line with this research theme, which is also the research agenda of other countries like Australia, Singapore and the Chinese Mainland.

"Developing a sustainable environment" means creating a sustainable and environmental-friendly development model for Hong Kong. Research on this theme may cover areas such as pollution prevention and control, sustainable energy, and recycling technology. This theme is of great importance because not only does it facilitate cooperation with the nearby regions and is ranked high on the research priority of governments and the academic community internationally, but it is also related to environmental industry which is one of the six priority economic areas identified by the Hong Kong Government.

Finally, the research theme "Strengthening Hong Kong's strategic position as a regional and international business center" can stand Hong Kong in a good stead to benefit from the rise of the Asian economy and enhance its status as a business center both regionally and internationally. Financial services and trading, logistics have always been two of the four traditional industries driving the Hong Kong economy. In light of the intense competition from the neighboring Asian cities, this research theme can contribute to Hong Kong's strategic economic development by strengthening our knowledge base in areas like risk management, wealth management and corporate governance, competitive models between Hong Kong and other Asian economies, and financial technology.

Professor Chin reiterates the importance of pooling together the best minds and resources to come up with the most beneficial solutions, such as engaging people from different academic domains to work on issues with major impact on Hong Kong. For instance, research on improving Hong Kong's air quality can involve researchers from many academic areas including mathematics, chemistry, engineering, and medicine, as well as academics in social sciences, law and economics. This can elevate the research impact of the Hong Kong academia to the next higher level.

Although the Scheme supports research of the UGC-funded tertiary institutions, overseas and Mainland scholars are encouraged to participate in the research projects. This will not only facilitate knowledge exchange between local and overseas researchers, but will encourage student participation from around the world, thereby enriching the research experience of the students studying in Hong Kong.

The RGC has called for white papers from local university researchers in early 2010, which will be used as a basis to identify the key research topics under each of the chosen themes. Once the key research topics have been identified, the RGC will invite proposals from Hong Kong researchers in September / October 2010. Successful research proposals will be awarded in 2011 after international peer review.

In: Biochemistry Researcher Biographical Sketches ... ISBN: 978-1-62081-308-9
Editors: J. T. Walker and C. M. Jones

Chapter 218

Chaos Theory in Astrophysics, Nature, Genetics, and Human Cancer

Jessica Li *
School of Biomedical Sciences,
Chinese University of Hong Kong, Shatin, N.T. Hong Kong

Research Summary

Chaos theory predicts the causation and possible outcomes of systems in nature and explains the ever increasing entropy of nature. By observing nonlinear data, it is utilized to find underlying order in apparently random data. Human cancer and diseases caused by aneuploidy, gene, and chromosome anomalies may be analyzed using this model as well as quantum mechanics and the universe as a whole. Chaos theory is applied from the very small world of the subatomic to the very large world of the cosmos. It can be used to observe many very different aspects of nature.

* Corresponding Author: Jessica Li, Email: jess.jsli@gmail.com Keywords: chaos theory, astrophysics, cancer, genetics.

In: Biochemistry Researcher Biographical Sketches ... ISBN: 978-1-62081-308-9
Editors: J. T. Walker and C. M. Jones

Chapter 219

Structural and Functional Studies of Zonula Occludens-1

Jia Chen*[*1,3], *Lifeng Pan*[1], *Jiang Yu*[1], *Zhiyi Wei*[1], *Yanxiang Zhao*[2] and *Mingjie Zhang*[1]

[1]Department of Biochemistry, Molecular Neuroscience Center,
Hong Kong University of Science and Technology, Clear Water Bay,
Kowloon, Hong Kong
[2]Department of Applied Biology and Chemical Technology,
Hong Kong Polytechnic University, Hong Ham, Kowloon, Hong Kong
[3]Department of Nanoscience and Nanotechnology,
Hong Kong University of Science and Technology, Clear Water Bay,
Kowloon, Hong Kong, China

Research Summary

Intercellular junctions, including tight junctions (TJs), adheren junctions (AJs), and gap junctions (GJs) as defence barrier, association points and communication channels plays important role in polarized epithelial cell or cell polarization events such as directional migration. Zonula occludens-1 (ZO-1), one of the junction marker proteins, is also the first identified member of the PDZ family of proteins. ZO family, ZO-1/2/3, are proved to have critical functions in the formation and maintenance of intercellular junctions. But how they organize and possibly regulate junctional complexes need to be further investigated. In my PG seminar talk, I will present our research progress in studying two regions of ZO-1: the second PDZ domain (PDZ2), and the multi-domain segment containing the third PDZ, the Src Homology domain (SH3), and the guanylate kinase (GuK)-like domain.

By using NMR, X-ray crystallography as structural approaches and biochemical methods such as isothermal titration, fluorescence polarity assay and etc, we found the PDZ2 domain of ZO-1 can form swapped dimer which generate a specific and regulatory

[*] Email: chenj@ust.hk.

Connexin43 (Cx43) binding groove. Another interesting finding about ZO-1 is that its third PDZ domain together with its following SH3 and GuK domains form a structural supramodule, which may provide functional information for its binding with TJ protein occludin, Y-box transcription factor ZONAB, and other junctional proteins. In summary, our studies have provided structural and functional indication on how ZO-1 modulates the junctional complexes and possible regulatory mechanism.

In: Biochemistry Researcher Biographical Sketches ... ISBN: 978-1-62081-308-9
Editors: J. T. Walker and C. M. Jones

Chapter 220

Global Gene Expression Changes during Chondrocyte Adaptation to ER Stress

***Zhijia Tan**[*1], **Ben Niu**[1], **Kwok Yeung Tsang**[1], **Ian Melhado**[1], **Michael Zhang**[2], **Danny Chan**[1] and **Kathryn S. E. Cheah**[1]*
[1]Department of Biochemistry, Faculty of Medicine, the University of Hong Kong
[2]Cold Spring Harbor Laboratory, NY, US

In humans, mutations in *COL10A1* trigger ER stress and result in metaphyseal chondrodysplasia type Schmid (MCDS). In the 13del transgenic mouse model with MCDS mutations, the Col10a1 transgene with a 13bp deletion is expressed in the hypertrophic chondrocytes (HCs). The mutant 13del proteins cannot fold properly in the endoplasmic reticulum (ER), thus inducing ER stress and trigger the unfolded protein response (UPR). The terminal differentiation of 13del HCs is impaired. These cells reprogram to alleviate ER stress and survive, including cell cycle re-entry and re-expression of markers characteristic of pre-hypertrophic chondrocytes (Tsang *et al*, 2007). To dissect the underlying mechanism of chondrocyte reprogramming and adaptation to stress, different regions in the wild type and 13del growth plates were fractionated for global gene expression profiling. Comparative transcriptomics between wild type and 13del HCs reveal novel ER stress inducible genes. Up regulation of genes expression related to oxidative stress and hypoxic stress pathways suggests that adaptation of HCs to ER stress in vivo involves cross- regulation by other stress pathways. Factors and regulatory circuitries that might be responsible for the chondrocyte reprogramming and ER stress adaptation will be predicted by bioinformatics approaches.

[*] E-mail: zjtan@hkusua.hku.hk.

Reference

Tsang, K. Y., Chan, D., Cheslett, D., Chan W. C. W., So C. L., Melhado, I. G., Chan, T. W. Y., Kwan K. M., Hunziker E. B., Yamada, Y. Bateman, J.F., Cheung K. M. C., Cheah, K.S.E. (2007). *PLoS Biology, 5*(3), 0569-0585.

In: Biochemistry Researcher Biographical Sketches … ISBN: 978-1-62081-308-9
Editors: J. T. Walker and C. M. Jones

Chapter 221

Genetic Analysis of Zebrafish Hematopoiesis-Defective Mutant Thy-193

Xiuling Li*[*1], Jin Xu[1], Yanmei Liu[2] and Zilong Wen[1]
[1]Department of Biochemistry, The Hong Kong University of Science and Technology, Clear Water Bay, Kowloon, P. R. China
[2]Max Planck Institute of Molecular Cell Biology and Genetics, Laboratory of Experimental Diabetology, Carl Gustav Carus Medical School, Dresden University of Technology, Dresden, Germany

The hematopoiesis in vertebrates includes two successive waves: 'primitive' and 'definitive' wave. Primitive hematopoiesis is a transient process and produces two blood cell lineages: myeloid (granulocytes and monocytes) and erythroid (erythrocytes and platelets) cells, whereas definitive wave, which is active throughout the lifetime of an organism, gives rise to all blood cells including erythrocytes, megakaryocytes (platelets), granulocyte, monocytes, and lymphocytes. Zebrafish has recently emerged as a powerful vertebrate model organism for study of hematopoiesis. Through forward genetic screen aimed to uncover zebrafish mutants with aberrant definitive hematopoiesis, a mutant zebrafish line Thy-193 has been identified.

By characterizing Thy-193 phenotype using whole mount in situ hybridization method, Thy-193 mutant shows normal primitive hematopoiesis, but aberrant maintenance of definitive hematopoietic stem cells(HSCs) without interfering the initiation of HSCs. By positional cloning, the Thy-193 mutant gene is mapped to sumolation activating enzyme subunit 1, sae1, which is a molecule involved in sumolation pathway. Sequencing of the sae1 gene shows a C817T mutation, resulting in a truncated protein lack of conserved C terminal 76 amino acids. This mapping result is further confirmed by rescue experiment with full-length sae1 RNA injection. Moreover, knocking down ubc9 to block sumolation pathway

* Email: xiuling@ust.hk

leads to a similar blood phenotype with Thy-193 mutant, implying sae1 might impact Thy-193 mutant via sumolation pathway.

In: Biochemistry Researcher Biographical Sketches ...
Editors: J. T. Walker and C. M. Jones
ISBN: 978-1-62081-308-9

Chapter 222

Structural and Functional Analysis of Influenza H5N1 Nucleoprotein Reveals Amino Acids That Are Crucial for Oligomerization, RNA Binding and Ribonucleoprotein Activities

A. K. L. Ng*[*1,2], W. H. Chan [1,2], N. C. Robb [4], M. K. H. Lam [1], P. K. S. Chan[3], S. W. N. Au [1,2], E. Fodor[4] and P. C. Shaw[1,2]

[1]Department of Biochemistry and Centre for Protein Science and Crystallography,
[2]Molecular Biotechnology Programme
[3]Department of Microbiology, Chinese University of Hong Kong,
Shatin, N.T., Hong Kong, China.
[4]Sir William Dunn School of Pathology, University of Oxford, South Parks Road,
Oxford, UK

The recent H1N1 influenza pandemic has become a major concern worldwide. The nucleoprotein (NP) of the virus binds the RNA genome and plays essential role in transcription and replication during the virus life cycle. We have determined the 3.3 Å crystal structure of H5N1 NP, which is composed of head and body domains and a tail loop. Using surface plasmon resonance, we found the basic loop (residues 73-91) and arginine-rich groove, but mostly a protruding element centering at R174 and R175, to be important in RNA binding by NP. To bind RNA, NP may first capture the RNA by the flexible basic loop and then the RNA is clamped by the protruding element. The crystal structure also revealed that NP forms homo-oligomers by inserting its distinctive tail loop into the body groove.

Using ribonucleoprotein (RNP) reconstitution assay, we identified E339, V408, P410, R416, L418 and L419 being crucial for forming a functional RNP, while R267, I406, R422 and E449 play supportive roles. Further characterization by static light scattering and co-

* Email: ng0567@gmail.com

immunoprecipitation showed that the reduction of RNP activities is due to the interference of NP homo-oligomerization.

We concluded that NP oligomerization event is governed by three factors: (1) interaction between the tail loop and the insertion groove; (2) maintenance of the tail loop conformation; and (3) stabilization of NP homo-oligomer. Our study of H5N1 NP sheds light on inhibitor design for combating influenza virus.

In: Biochemistry Researcher Biographical Sketches ... ISBN: 978-1-62081-308-9
Editors: J. T. Walker and C. M. Jones

Chapter 223

Structural Basis of the Organization of the USH1 Interactome

Lifeng Pan*, Jing Yan, Lin Wu and Mingjie Zhang
Department of Biochemistry, Hong Kong University of Science and Technology, Clear Water Bay, Kowloon, Hong Kong, China

The hereditary hearing-vision loss disease Usher syndrome I (USH1) is caused by defects in several proteins, which form an integrated protein network called the USH1 interactome. The proper assembly of the USH1 interactome is essential for both the development and maintenance of visual and hair cells. However, the biochemical and structural mechanism governing the USH1 protein complex formation is largely unclear. Using NMR spectroscopy, we first elucidate the multi-dentate binding mode between harmonin and cadherin23, which provides a structural and biochemical basis for the harmonin-mediated assembly of stable tip link complex in the auditory hair cells. Then using X-ray Crystallography combined with NMR, we solved the high resolution structure of the nucleation core of the USH1 complex formed by harmonin and Sans. The complex structure not only reveals an unexpected assembly mode between harmonin and Sans but also uncovers a previously unknown interaction mode for both PDZ and SAM domains. The synergistic PDZ1/SAM and PDZ1/carboxyl PDZ binding-motif interactions between harmonin and Sans lock the two scaffold proteins into a highly stable complex, acting as the nucleation core of the USH1 interactome. In summary, our findings provide the first atomic picture showing the assembly of the USH1 interactome.

In addition to providing molecular explanations to the disease-causing mutations already found in USH1 patients, the complex structures will undoubtedly be valuable in guiding future discoveries of other mutations in these USH1 proteins in human patients.

* E-mail: panly@ust.hk